浙江省高等教育重点建设教材

单片机实验与课程设计指导
（Proteus仿真版）
（第二版）

楼然苗	胡佳文	李光飞	
刘玉良	李韵磊	叶继英	编著

U0277151

ZHEJIANG UNIVERSITY PRESS

浙江大学出版社

内容简介

单片机实验与课程设计指导是学生加深理论知识理解、提高实际设计能力的重要环节,从学生自己设计电路板,到程序编制与调试,最后完成一个单片机系统的设计,可以让学生体验成功的快乐! proteus 虚拟单片机仿真软件可以成功地进行绝大部分的单片机硬件仿真,从而在教师进行课堂教学或实验设计、演示等环节,可以轻松实现程序功能的展示。本书在原《单片机实验与课程设计指导(proteus 仿真版)》教材的基础上增加了新的实验与课程设计内容,适合教师在单片机课程教学中进行教学程序功能演示及作为学生实验与课程设计的指导用书。

图书在版编目（CIP）数据

单片机实验与课程设计指导:Proteus 仿真版 / 楼
然苗等编著. —2 版. —杭州:浙江大学出版社，2013. 7(2019.1重印)
 ISBN 978-7-308-11906-1

Ⅰ. ①单… Ⅱ. ①楼… Ⅲ. ①单片微型计算机－系统
设计－应用软件－高等学校－教学参考资料　Ⅳ.
①TP368.1

中国版本图书馆 CIP 数据核字（2013）第 170919 号

单片机实验与课程设计指导(Proteus 仿真版)(第二版)

楼然苗	胡佳文	李光飞	
刘玉良	李韵磊	叶继英	**编著**

责任编辑	吴昌雷	
封面设计	刘依群	
出版发行	浙江大学出版社	
	（杭州市天目山路 148 号　邮政编码 310007）	
	（网址:http://www.zjupress.com)	
排　版	杭州中大图文设计有限公司	
印　刷	虎彩印艺股份有限公司	
开　本	787mm×1092mm　1/16	
印　张	15.75	
字　数	383 千	
版印次	2013 年 7 月第 2 版　2019 年 1 月第 4 次印刷	
书　号	ISBN 978-7-308-11906-1	
定　价	32.00 元	

前　言

　　Proteus 是一种功能强大的电子设计自动化软件,提供智能原理图设计系统,能模拟数字电路、模拟电路及 MCU 器件混合仿真系统和 PCB 设计系统功能。Proteus 软件不仅可以仿真传统的电路分析实验、模拟电子线路实验、数字电路实验,而且可以提供嵌入式系统(单片机应用系统、ARM 应用系统)的仿真实验。它支持单片机和周边设备,可以仿真 51 系列、AVR、PIC、Motorola 的 68 系列等常用的 MCU,并可提供周边设备的仿真,例如字符 LCD 模块、图形 LCD 模块、LED 点阵、LED 七段显示模块、键盘/按键、直流/步进/伺服电机、RS232 虚拟终端、电子温度计、示波器等。Proteus 提供了大量的元件库,有 RAM、ROM、键盘、马达、LED、LCD、AD/DA、部分 SPI 器件、部分 IIC 器件等。在编译方面,它支持单片机汇编语言的编辑/编译/源码级仿真,也支持 Keil 和 MPLAB 等多种编译器。内带的 8051、AVR、PIC 的汇编编译器,也可以与第三方集成编译环境(如 IAR、Keil 和 Hitech)结合,进行高级语言的源码级仿真和调试。

　　利用 Proteus 的单片机硬件电路进行程序运行效果仿真,可以方便直观地进行单片机程序运行效果演示,极大地拓展了课堂教学及实验教学的硬件环境条件,老师或学生可以在教室或寝室方便地利用电脑进行单片机程序的调试及效果演示,为设计开发单片机应用产品提高了效率。

　　本书是用于教师或学生进行单片机实验或课程设计的指导书。本书中的所有实验程序及设计硬件电路资料可在浙江海洋学院精品课程网站(http://61.153.216.116/jpkc/jpkc/dpj/News/Show.asp? id＝60＆cataid＝A0049)和浙江大学出版社网站(www.zjupress.com)中获得,以方便学校老师及学生教学与学习使用。用于 Proteus 仿真的单片机程序都可以在真实硬件电路板上运行,为教师课堂教学实验的演示或设计程序的功能演示提供了极大的方便。

　　本书选用 89C52 系列单片机作为处理器,选择了课堂教学或课程实验中,以定时器使用、中断使用、串行口使用、七段 LED 显示器动态扫描显示、LCD 点阵液晶显示,以及在课程设计中对彩灯控制器设计、单片机时钟设计、DS18B20 数字温度计设计、DS1302 实时时钟设计、低频信号发生器设计、电子密码锁等例子,较详细地介绍了系统功能、设计方案、硬件仿真电路、程序设计、仿真运行结果等,书中的源程序及电路图可供参考。

　　本书 Proteus 仿真电路图设计采用 Proteus 7.1 版本,使用时请安装 Proteus 7.1 及以上版本仿真软件。有关 Proteus 仿真软件的安装与使用方法请参考相关资料,本书中不进行相关介绍。

　　感谢浙江大学出版社在本书出版过程中给予的帮助与支持!

　　作者邮箱:louranmiao@zjou.edu.cn

作　者

2013 年 3 月

目　　录

第 1 章

实验一：LED 小灯实验

1.1 实验内容与要求

1. 实验目的

（1）学习用程序延时的方法进行 LED 小灯的亮灭控制。

（2）学习掌握流水小灯的编程方法。

2. Proteus 仿真实验硬件电路

LED 小灯实验的 Proteus 仿真实验硬件电路如图 1.1 所示。

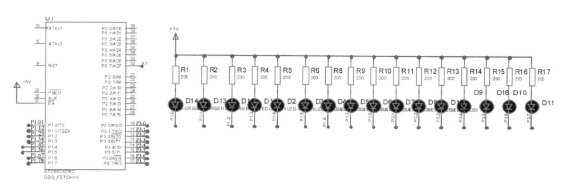

图 1.1　LED 小灯实验的仿真硬件电路

3. 实验任务

完成对接在 P1、P3 端口的发光二极管闪亮控制程序的设计和调试。具体要求：

（1）用程序延时的方法让 P1 口的 1 个 LED 小灯每隔 1s 交替闪亮。

(2)用程序延时的方法让 P1 口的 8 个 LED 发光二极管循环闪亮(每个亮 50ms)。

(3)用程序延时的方法让 P1 口的 8 个 LED 小灯追逐闪亮(50ms 间隔变化)。

(4)用程序延时的方法让 P1、P3 口的 16 个 LED 小灯循环闪亮(每个亮 50ms)。

4. 实验预习要求

(1)根据硬件电路原理图,分析二极管点亮的条件;复习延时子程序中延时时间的计算方法,会计算延时子程序的初值。

(2)根据硬件电路原理图,画出实际接线图。

(3)根据实验任务设计出相应的调试程序。

(4)学习掌握 Wave、Madwin、Keil-51 等编译软件的使用方法。

(5)完成预习报告。

5. 实验设备

计算机(已安装单片机汇编编译软件及 Proteus 软件)。

6. 实验报告要求

整理好实验任务 1~4 中经 Proteus 运行正确的程序。

1.2　实验一参考汇编程序

```
;**********************************************************;
;                       实验程序 1.1                      ;
;     用程序延时的方法让 P1 口的 1 个 LED 小灯每隔 1s 交替闪亮   ;
;                       12MHz 晶振                         ;
;**********************************************************;
        ORG    0000H     ;程序执行开始地址
        LJMP   START     ;跳至 START 执行
;
        ORG    0030H     ;以下程序放在 0030H 地址后
START: CPL    P1.0
        LCALL  DL1S
        AJMP   START
;
;约 0.5ms 延时子程序,执行一次时间为 503 μs
DL503: MOV    R2,#250
LOOP1: DJNZ   R2,LOOP1
```

```
        RET
;
;约 10ms 延时子程序(调用 20 次 0.5ms 延时子程序)
DL10ms:MOV    R3,#20
LOOP2: LCALL   DL503
        DJNZ    R3,LOOP2
        RET
;
;约 1s 延时子程序
DL1S:  MOV     R4,#100
LOOP3: LCALL   DL10ms
        DJNZ    R4,LOOP3
        RET
;
END             ;  结束
```

实验程序 1.1 的 Proteus 仿真效果如图 1.2 所示。

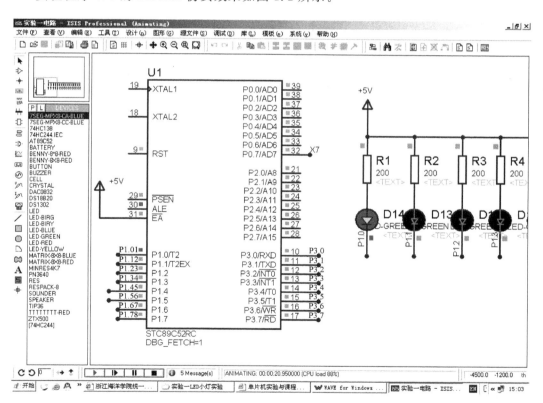

图 1.2 实验程序 1.1 的 Proteus 仿真效果

```
;**********************************************************;
;                        实验程序 1.2                      ;
;用程序延时的方法让 P1 口的 8 个 LED 发光二极管循环闪亮(每个亮 50ms);
;                        12MHz 晶振                        ;
;**********************************************************;
        ORG     0000H       ;程序执行开始地址
        LJMP    START       ;跳至 START 执行
;
        ORG     0030H       ;以下程序放在 0030H 地址后
START:  LCALL   FUN0
        AJMP    START
;
;循环闪亮功能子程序
FUN0:   MOV     A,#0FEH         ;累加器赋初值
FUN00:  MOV     P1,A            ;累加器值送至 P1 口
        LCALL   DL50ms          ;延时
        JNB     ACC.7,OUT       ;累加器最高位为 0 时结束
        RL      A               ;累加器 A 中数据循环左移 1 位
        AJMP    FUN00           ;转 FUN00 循环
OUT:    RET
;
;约 0.5ms 延时子程序,执行一次时间为 503 μs
DL503:  MOV     R2,#250
LOOP1:  DJNZ    R2,LOOP1
        RET
;
;约 50ms 延时子程序(调用 100 次 0.5ms 延时子程序)
DL50ms: MOV     R3,#100
LOOP2:  LCALL   DL503
        DJNZ    R3,LOOP2
        RET
;
END
```

实验程序 1.2 的 Proteus 仿真效果如图 1.3 所示。

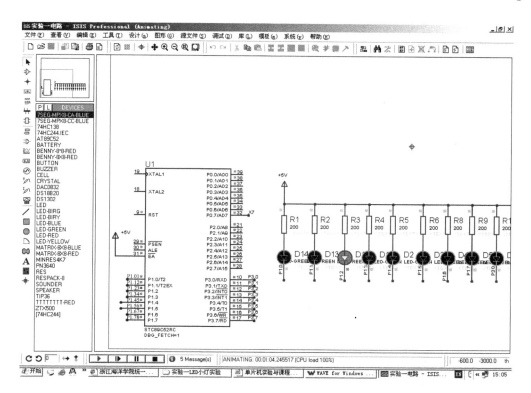

图 1.3　实验程序 1.2 的 Proteus 仿真效果

```
;*******************************************************;
;                    实验程序 1.3                      ;
;用程序延时的方法让 P1 口的 8 个 LED 小灯追逐闪亮(50ms 间隔变化);
;                     12MHz 晶振                        ;
;*******************************************************;
       ORG    0000H      ;程序执行开始地址
       LJMP   START      ;跳至 START 执行
;
       ORG    0030H      ;以下程序放在 0030H 地址后
START: LCALL  FUN1
       AJMP   START
;
;追逐闪亮功能子程序
FUN1:  MOV    A,#0FEH     ;累加器赋初值
FUN11: MOV    P1,A        ;累加器值送至 P1 口
       LCALL  DL50ms      ;延时
       JZ     OUT         ;A 为 0 结束
       RL     A           ;累加器 A 中数据循环左移 1 位
```

```
        ANL     A,P1              ;A 同 P1 口值相与
        AJMP    FUN11             ;转 FUN11 循环
OUT:    RET
;
;约 0.5ms 延时子程序,执行一次时间为 503 μs
DL503:  MOV     R2,♯250
LOOP1:  DJNZ    R2,LOOP1
        RET
;约 50ms 延时子程序(调用 100 次 0.5ms 延时子程序)
DL50ms: MOV     R3,♯100
LOOP2:  LCALL   DL503
        DJNZ    R3,LOOP2
        RET
;
END
```

实验程序 1.3 的 Proteus 仿真效果如图 1.4 所示。

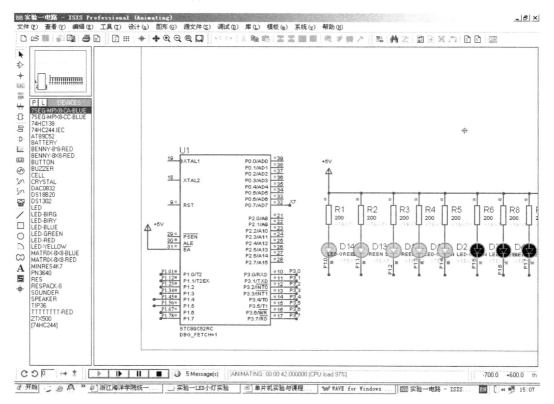

图 1.4　实验程序 1.3 的 Proteus 仿真效果

```
;**********************************************************;
;                    实验程序 1.4                        ;
; 用程序延时的方法让 P1、P3 口的 16 个 LED 小灯循环闪亮(每个亮 50ms) ;
;                    12MHz 晶振                           ;
;**********************************************************;
        ORG     0000H       ;程序执行开始地址
        LJMP    START       ;跳至 START 执行
;
        ORG     0030H       ;以下程序放在 0030H 地址后
START：  LCALL   FUNP1
        MOV     P1,♯0FFH
        LCALL   FUNP3
        MOV     P3,♯0FFH
        AJMP    START
;
;循环闪亮功能子程序
FUNP1：  MOV     A,♯0FEH     ;累加器赋初值
FUN11：  MOV     P1,A        ;累加器值送至 P1 口
        LCALL   DL50ms      ;延时
        JNB     ACC.7,OUT   ;累加器最高位为 0 时结束
        RL      A           ;累加器 A 中数据循环左移 1 位
        AJMP    FUN11       ;转 FUN11 循环
OUT：    RET
FUNP3：  MOV     A,♯0FEH     ;累加器赋初值
FUN33：  MOV     P3,A        ;累加器值送至 P3 口
        LCALL   DL50ms      ;延时
        JNB     ACC.7,OUT   ;累加器最高位为 0 时结束
        RL      A           ;累加器 A 中数据循环左移 1 位
        AJMP    FUN33       ;转 FUN33 循环
;
;约 0.5ms 延时子程序,执行一次时间为 503μs
DL503：  MOV     R2,♯250
LOOP1：  DJNZ    R2,LOOP1
        RET
;
;约 50ms 延时子程序(调用 100 次 0.5ms 延时子程序)
DL50ms：MOV     R3,♯100
LOOP2： LCALL   DL503
        DJNZ    R3,LOOP2
        RET
;
        END
```

実験程序 1.4 的 Proteus 仿真効果如図 1.5 所示。

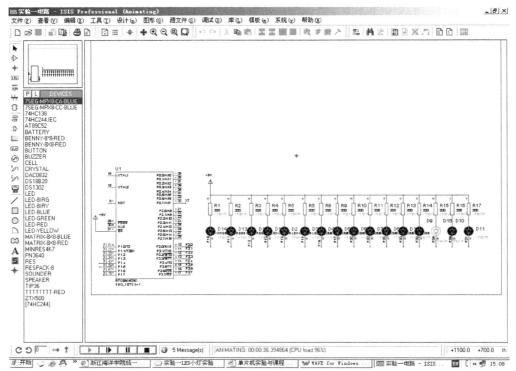

图 1.5　実験程序 1.4 的 Proteus 仿真効果

1.3　実験一参考 C 程序

```
/*-------------------------------------
LED program V1.1
MCU STC89C52RC   XAL 12MHz
Build by Gavin Hu，2010.6.1
------------------------------------- */
# include <reg51.h>
sbit P11 = P1^1;
void delay_ms(unsigned int);

/*-------------------------------------
main function
------------------------------------- */
```

```
void main(void)
{
while(1)
    {
    P11 = ! P11;
    delay_ms(1000);
    }
}


/*------------------------------------
  Delay function
  Parameter: unsigned int dt
  Delay time = dt(ms)
------------------------------------ */
void delay_ms(unsigned int dt)
{
register unsigned char bt,ct;
for (;dt;dt -- )
    for (ct = 2;ct;ct -- )
        for (bt = 250; -- bt;);
}


/*------------------------------------
LED program V1. 2
MCU STC89C52RC   XAL 12MHz
Build by Gavin Hu，2010. 6. 1
------------------------------------ */
# include <reg51. h>
void delay_ms(unsigned int);


/*------------------------------------
  main function
------------------------------------ */
void main(void)
{
P1 = 0xfe;
while(1)
    {
    P1 = (P1<<1)|(P1>>7);
```

```
        delay_ms(50);
    }
}


/*-----------------------------------
  Delay function
  Parameter: unsigned int dt
  Delay time = dt(ms)
----------------------------------- */
void delay_ms(unsigned int dt)
{
register unsigned char bt,ct;
for (;dt;dt --)
    for (ct = 2;ct;ct --)
        for (bt = 250; -- bt;);
}


/*-----------------------------------
LED program V1.3
MCU STC89C52RC   XAL 12MHz
Build by Gavin Hu, 2010.6.1
----------------------------------- */
# include <reg51.h>
void delay_ms(unsigned int);


/*-----------------------------------
  main function
----------------------------------- */
void main(void)
{
unsigned char i;
while(1)
    {
    P1 = 0xff;
    for (i = 0;i<8;i ++ )
        {
        P1 = P1<<1;
        delay_ms(50);
        }
```

```
            }
    }


/*------------------------------------
   Delay function
   Parameter: unsigned int dt
   Delay time = dt(ms)
   ------------------------------------ */
void delay_ms(unsigned int dt)
{
register unsigned char bt,ct;
for (;dt;dt --)
    for (ct = 2;ct;ct --)
        for (bt = 250; -- bt;);
}


/*------------------------------------
LED program V1. 4
MCU STC89C52RC   XAL 12MHz
Build by Gavin Hu, 2010. 6. 1
   ------------------------------------ */
# include <reg51. h>
void delay_ms(unsigned int);


/*------------------------------------
   main function
   ------------------------------------ */
void main(void)
{
P1 = 0xfe;
P3 = 0xff;
while(1)
    {
    delay_ms(50);
    P1 = (P1<<1)|(P3>>7);
    P3 = (P3<<1)|(P1>>7);
    }
}
```

```
/*---------------------------------------
  Delay function
  Parameter: unsigned int dt
  Delay time = dt(ms)
--------------------------------------- */
void delay_ms(unsigned int dt)
{
register unsigned char bt,ct;
for (;dt;dt --)
    for (ct = 2;ct;ct --)
        for (bt = 250; -- bt;);
}
```

第 2 章

实验二:定时/计数器实验

2.1 实验内容与要求

1. 实验目的

(1)学习掌握定时/计数器程序初始化的设计方法。

(2)学习掌握定时/计数器方式 1、方式 2 的使用编程方法。

2. Proteus 仿真实验硬件电路

定时/计数器实验的 Proteus 仿真实验硬件电路如图 2.1 所示。

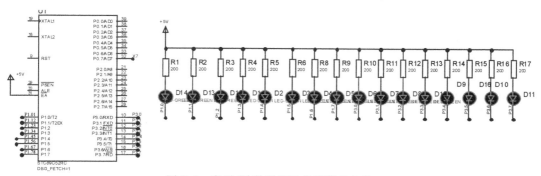

图 2.1 定时/计数器实验仿真硬件电路

3. 实验任务

完成对接在 P1、P3 端口的发光二极管闪亮控制程序的设计和调试。具体要求:

(1)选择定时器 T0 为工作方式 1,定时溢出时间为 50ms,使 P1 口的 8 个发光二极管循环闪亮。

(2)选择定时器 T0 为工作方式 1,定时溢出时间为 50ms,使 P1.0 口的 1 个发光二极管每隔 1s 交替闪亮。

(3)使用定时器 T0、T1 为工作方式 1,定时溢出时间为 50ms,分别控制 P1、P3 口的小灯,使其对应端口的 8 个发光二极管循环闪亮。

(4)将 T0 定时器设定为工作方式 2,使 P1.0 口的 1 个发光二极管每隔 50ms 交替闪亮。

4. 实验预习要求

(1)根据硬件电路原理图,分析 LED 发光二极管点亮的条件,画出实际接线图。

(2)阅读教材中有关定时/计数器的内容,熟悉定时/计数器的基本结构和工作过程;计算 50ms 定时/计数器时间常数;根据实验任务设计出相应的调试程序。

(3)掌握 Wave、Madwin、Keil-51 等编译软件的使用方法。

(4)完成预习报告。

5. 实验设备

计算机(已安装单片机汇编编译软件及 Proteus 软件)。

6. 思考题

定时器工作于方式 1、方式 2 时,其一次溢出的最大定时时间是多少(设单片机的晶振为 12MHz)?

7. 实验报告要求

(1)整理好实验任务 1~4 中经 Proteus 运行正确的程序。

(2)解答思考题。

2.2   实验二参考汇编程序

```
;*****************************************************;
;                      实验程序 2.1                  ;
;    选择定时器 T0 为工作方式 1,定时溢出时间为 50ms    ;
;    使 P1 口的 8 个发光二极管循环闪亮                 ;
;*****************************************************;
        ORG    0000H
        LJMP   MAIN
;
MAIN:   MOV    TMOD,#11H      ; 设 T0、T1 为 16 位定时器模式
        MOV    TL0,#0B0H      ; 赋 50ms 初值
        MOV    TH0,#3CH       ; 赋 50ms 初值
        MOV    P1,#11111110B  ; 预置 P1 口小灯控制初值
```

```
            SETB    TR0             ; 开启定时器 T0
    LOOP:   JBC TF0,CPLP            ; TF0 为 1,转 CPLP 并将 TF0 清 0
            AJMP    LOOP            ; TF0 为 0,则转 LOOP 循环等待
    CPLP:   MOV     TL0,#0B0H       ; T0 重装初值
            MOV     TH0,#3CH        ;
            MOV     A,P1            ; 将端口 P1 中值读入 A 中
            RL      A               ; A 中二进制数循环左移
            MOV     P1,A            ; 控制 P1 端口小灯状态
            AJMP    LOOP            ; 转 LOOP 再循环等待 50ms
    ;
            END                     ; 结束
```

实验程序 2.1 的 Proteus 仿真效果如图 2.2 所示。

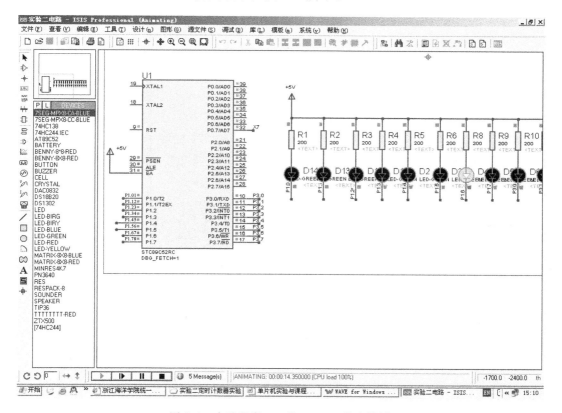

图 2.2 实验程序 2.1 的 Proteus 仿真效果

```
;************************************************************;
;                      实验程序 2.2                        ;
;         定时器 T0 为工作方式 1,定时溢出时间为 50ms         ;
;          使 P1.0 口的 1 个发光二极管每隔 1s 交替闪亮        ;
;************************************************************;
```

```
        ORG     0000H
        LJMP    MAIN
;
MAIN:   MOV     TMOD,＃01H        ; 设 T0 为 16 位定时器模式
        MOV     TL0,＃0B0H        ; 赋 50ms 初值
        MOV     TH0,＃3CH         ; 赋 50ms 初值
        MOV     R0,＃20           ; 预置定时控制值(50ms×20＝1s)
        SETB    TR0              ; 开启定时器 T0
LOOP:   JBC     TF0,CPLP         ; TF0 为 1,转 CPLP 并将 TF0 清 0
        AJMP    LOOP             ; TF0 为 0,则转 LOOP 循环等待
CPLP:   MOV     TL0,＃0B0H        ; 重装初值
        MOV     TH0,＃3CH         ;
        DJNZ    R0,LOOP          ; 判断是否到 20 次溢出时间
        MOV     R0,＃20           ; 重装预置定时控制值
        CPL     P1.0             ; 改变 P 小灯亮灭状态
        AJMP    LOOP             ; 转 LOOP 再循环等待
        END                      ; 结束
```

实验程序 2.2 的 Proteus 仿真效果如图 2.3 所示。

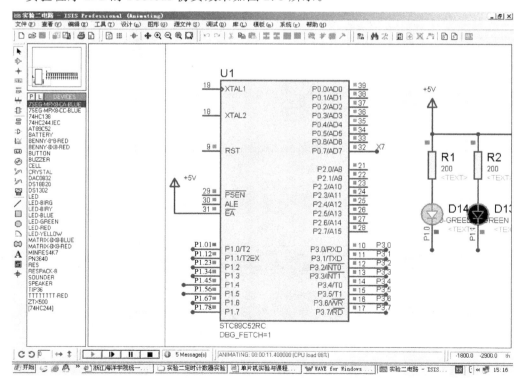

图 2.3   实验程序 2.2 的 Proteus 仿真效果

```
;**********************************************************;
;                    实验程序 2.3                         ;
;    定时器 T0、T1 为工作方式 1,定时溢出时间为 50ms,分别控制    ;
;    P1、P3 口的小灯,使各对应端口的 8 个发光二极管循环闪亮      ;
;**********************************************************;
        ORG     0000H
        LJMP    MAIN
;
MAIN:   MOV     TMOD,♯11H        ; 设 T0、T1 为 16 位定时器模式
        MOV     TL0,♯0B0H        ; 赋 50ms 初值
        MOV     TH0,♯3CH         ; 赋 50ms 初值
        MOV     TL1,♯0B0H        ; 赋 50ms 初值
        MOV     TH1,♯3CH         ; 赋 50ms 初值
        MOV     P1,♯11111110B    ; 预置 P1 口小灯控制初值
        MOV     P3,♯01111111B    ; 预置 P3 口小灯控制初值
        SETB    TR0              ; 开启定时器 T0
        SETB    TR1              ; 开启定时器 T1
LOOP:   JBC     TF0,CPLP         ; TF0 为 1,转 CPLP 并将 TF0 清 0
        JBC     TF1,CPLP1        ; TF1 为 1,转 CPLP1 并将 TF1 清 0
        AJMP    LOOP             ; 转 LOOP 循环等待
CPLP:   MOV     TL0,♯0B0H        ; T0 重装初值
        MOV     TH0,♯3CH         ;
        MOV     A,P1             ; 将端口 P1 中值读入 A 中
        RL      A                ; A 中二进制数循环左移
        MOV     P1,A             ; 控制 P1 端口小灯状态
        AJMP    LOOP             ; 转 LOOP 再循环等待 50ms
;
CPLP1:  MOV     TL1,♯0B0H        ; T1 重装初值
        MOV     TH1,♯3CH         ;
        MOV     A,P3             ; 将端口 P3 中值读入 A 中
        RR      A                ; A 中二进制数循环右移
        MOV     P3,A             ; 控制 P3 端口小灯状态
        AJMP    LOOP             ; 转 LOOP 再循环等待 50ms
        END                      ; 结束
```

实验程序 2.3 的 Proteus 仿真效果如图 2.4 所示。

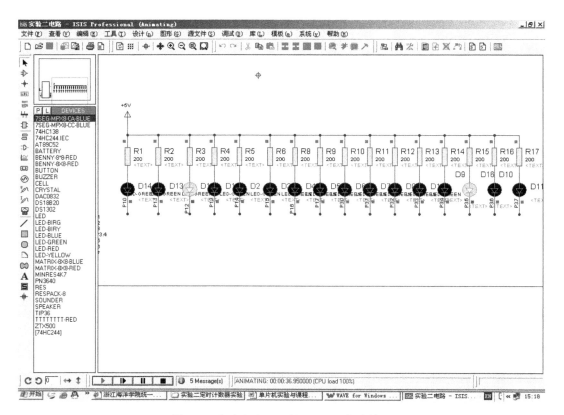

图 2.4   实验程序 2.3 的 Proteus 仿真效果

```
;*****************************************************************;
;                      实验程序 2.4                              ;
;    定时器 T0 为工作方式 2                                       ;
;    使 P1.0 口的 1 个发光二极管每隔 50ms 交替闪亮                ;
;*****************************************************************;
        ORG     0000H
        LJMP    MAIN
;
MAIN:   MOV     TMOD,#02H      ; 设 T0 为 8 位自动重装定时器模式
        MOV     TL0,#06H       ; 赋 250 μs 初值
        MOV     TH0,#06H       ; 赋 250 μs 初值
        MOV     R0,#200        ; 预置定时控制值(200×250 μs = 50ms)
        SETB    TR0            ; 开启定时器 T0
LOOP:   JBC     TF0,CPLP       ; TF0 为 1,转 CPLP 并将 TF0 清 0
        AJMP    LOOP           ; TF0 为 0,则转 LOOP 循环等待
CPLP:   DJNZ    R0,LOOP        ; 判断是否到 200 次溢出时间
```

```
MOV      R0，♯200            ;  重装预置定时控制值
CPL      P1.0                ;  改变 P1.0 小灯亮灭状态
AJMP     LOOP                ;  转 LOOP 再循环
END                         ;  结束
```

　　实验程序 2.4 的 Proteus 仿真效果如图 2.5 所示。

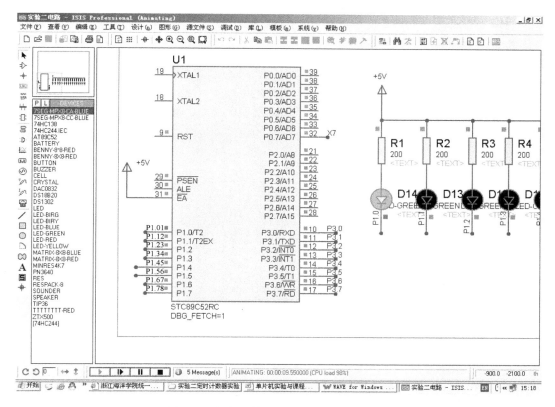

图 2.5　实验程序 2.4 的 Proteus 仿真效果

2.3　实验二参考 C 程序

```
/*------------------------------------
Timer program V2.1
MCU STC89C52RC   XAL 12MHz
Build by Gavin Hu，2010.6.1
------------------------------------ */
♯include <reg51.h>
```

```c
/*--------------------------------------
   main function
-------------------------------------- */
void main(void)
{
TMOD = 0x01;
TH0 = 0x3C;
TL0 = 0xB0;
P1 = 0xfe;
TR0 = 1;
while(1)
    {
    while(!TF0);
    TL0 = 0xB0;
    TH0 = 0x3C;
    TF0 = 0;
    P1 = (P1<<1)|(P1>>7);
    }
}


/*--------------------------------------
Timer program V2.2
MCU STC89C52RC   XAL 12MHz
Build by Gavin Hu, 2010.6.1
-------------------------------------- */
#include <reg51.h>
sbit P10 = P1^0;
/*--------------------------------------
   main function
-------------------------------------- */
void main(void)
{
unsigned char i;
TMOD = 0x01;
TH0 = 0x3C;
TL0 = 0xB0;
TR0 = 1;
while(1)
    {
```

```c
        P10 = ! P10;
        for ( i = 0;i<20;i ++ )
            {
            while( ! TF0);
            TL0 = 0xB0;
            TH0 = 0x3C;
            TF0 = 0;
            }
        }
}


/*-------------------------------------
Timer program V2. 3
MCU STC89C52RC   XAL 12MHz
Build by Gavin Hu, 2010. 6. 1
------------------------------------- */
# include <reg51.h>


/*-------------------------------------
  main function
------------------------------------- */
void main(void)
{
TMOD = 0x11;
TH0 = 0x3C;
TL0 = 0xB0;
TH1 = 0x3C;
TL1 = 0xB0;
P1 = 0x7f;
P3 = 0xfe;
TR0 = 1;
TR1 = 1;
while(1)
    {
    if (TF0)
        {
        TL0 = 0xB0;
        TH0 = 0x3C;
        TF0 = 0;
```

```
            P1 = (P1>>1)|(P1<<7);
        }
    if (TF1)
        {
        TL1 = 0xB0;
        TH1 = 0x3C;
        TF1 = 0;
        P3 = (P3<<1)|(P3>>7);
        }
    }
}


/*-------------------------------------
Timer program V2.4
MCU STC89C52RC   XAL 12MHz
Build by Gavin Hu, 2010.6.1
------------------------------------- */
# include <reg51.h>
sbit P10 = P1^0;


/*-------------------------------------
  main function
------------------------------------- */
void main(void)
{
unsigned char i;
TMOD = 0x02;
TH0 = 0x06;
TL0 = 0x06;
TR0 = 1;
while(1)
    {
    P10 = ! P10;
    for (i = 0;i<200;i++)
        {
        while(! TF0);
        TF0 = 0;
        }
    }
}
```

第 3 章

实验三：定时器中断实验

3.1 实验内容与要求

1. 实验目的

(1) 学习掌握定时器中断程序初始化的设计方法。
(2) 学习掌握定时器中断程序的编程方法。

2. Proteus 仿真实验硬件电路

定时器中断实验的 Proteus 仿真实验硬件电路如图 3.1 所示。

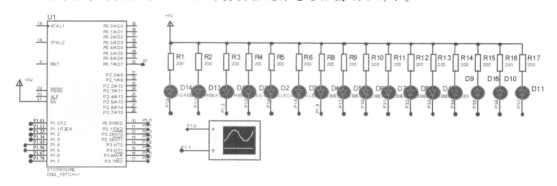

图 3.1　定时器中断实验的仿真硬件电路

3. 实验任务

在 P1 端口输出一定周期要求的方波信号或小灯亮灭控制信号。具体要求：
(1) 利用 T0 的定时中断法，在 P1.0 端口产生 500Hz(周期 2ms)的对称方波脉冲。

(2)利用 T0、T1 定时中断,在 P1.0 端口与 P1.1 端口分别产生 500Hz(周期 2ms)、1000Hz(周期 1ms)的对称方波脉冲。

4. 实验预习要求

(1)根据硬件电路原理图,画出实际接线图。

(2)阅读教材中有关定时器中断的内容,熟悉定时器中断的基本程序结构和工作过程;计算 500Hz(周期 2ms)、1000Hz(周期 1ms)的对称方波脉冲所需定时器初值;根据实验任务设计出相应的调试程序。

(3)掌握 Wave、Madwin、Keil-51 等编译软件的使用方法。

(4)完成预习报告。

5. 实验设备

计算机(已安装单片机汇编编译软件及 Proteus 软件)。

6. 思考题

怎样用定时器中断的方法实现 P1 口或 P3 口的 LED 小灯循环闪亮或追逐闪亮?

7. 实验报告要求

(1)整理好实验任务 1~2 中经 Proteus 运行正确的程序。

(2)解答思考题并进行程序的汇编与仿真调试,整理好经 Proteus 运行正确的思考题程序。

3.2　实验三参考汇编程序

```
;**********************************************************;
;                      实验程序 3.1                       ;
;   利用 T0 的定时中断法,在 P1.0 端口产生 500Hz 的对称方波脉冲    ;
;                      12MHz 晶振                          ;
;**********************************************************;
;
        ORG     0000H           ;主程序执行入口地址
        LJMP    MAIN            ;跳至 MAIN 执行
        ORG     000BH           ;T0 溢出中断服务程序入口
        LJMP    INTT0           ;跳至 T0 溢出中断服务程序
MAIN:   MOV     TMOD,#01H       ;T0 为 16 位定时模式
        MOV     TL0,#18H        ;定时器装初值(溢出时间 1ms)
```

```
        MOV     TH0,♯0FCH              ;定时器装初值
        SETB    EA                     ;开总中断允许
        SETB    ET0                    ;开定时器 T0 中断允许
        SETB    TR0                    ;开启定时器 T0
        SJMP    $                      ;等待
INTT0:  CPL     P1.0                   ;P1.0 取反
        MOV     TL0,♯18H               ;重装初值
        MOV     TH0,♯0FCH              ;重装初值
        RETI                           ;中断返回
        END                            ;结束
```

实验程序 3.1 的 Proteus 仿真效果如图 3.2 所示。

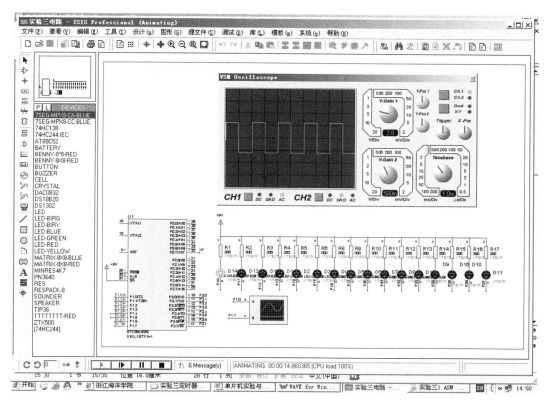

图 3.2　实验程序 3.1 的 Proteus 仿真效果

```
;********************************************************** ;
;                      实验程序 3.2                         ;
;  利用 T0、T1 定时中断,在 P1.0 端口与 P1.1 端口分别产       ;
;  生 500Hz(周期 2ms)、1000Hz(周期 1ms)的对称方波脉冲        ;
;                      12MHz 晶振                            ;
;********************************************************** ;
```

```
;
        ORG     0000H               ; 主程序执行入口地址
        LJMP    MAIN                ; 跳至 MAIN 执行
        ORG     000BH               ; T0 溢出中断服务程序入口
        LJMP    INTT0               ; 跳至 T0 溢出中断服务程序
        ORG     001BH               ; T1 溢出中断服务程序入口
        LJMP    INTT1               ; 跳至 T1 溢出中断服务程序
MAIN:   MOV     TMOD,#11H           ; T0、T1 为 16 位定时模式
        MOV     TL0,#18H            ; 定时器装初值(溢出时间 1ms)
        MOV     TH0,#0FCH           ; 定时器装初值
        MOV     TL1,#0CH            ; 定时器装初值(溢出时间 0.5ms)
        MOV     TH1,#0FEH           ; 定时器装初值
        SETB    EA                  ; 开总中断允许
        SETB    ET0                 ; 开定时器 T0 中断允许
        SETB    ET1                 ; 开定时器 T1 中断允许
        SETB    TR0                 ; 开启定时器 T0
        SETB    TR1                 ; 开启定时器 T1
        SJMP    $                   ; 等待
INTT0:  CPL     P1.0                ; P1.0 取反
        MOV     TL0,#18H            ; T0 重装初值
        MOV     TH0,#0FCH           ; T0 重装初值
        RETI                        ; 中断返回
INTT1:  CPL     P1.1                ; P1.1 取反
        MOV     TL1,#0CH            ; T1 重装初值
        MOV     TH1,#0FEH           ; T1 重装初值
        RETI                        ; 中断返回
        END                         ; 结束
```

实验程序 3.2 的 Proteus 仿真效果如图 3.3 所示。

3.3 实验三参考 C 程序

```
/*-------------------------------------
Timer interrupt program V3.1
MCU STC89C52RC  XAL 12MHz
Build by Gavin Hu, 2010.6.1
--------------------------------------- */
# include <reg51.h>
sbit P10 = P1^0;
```

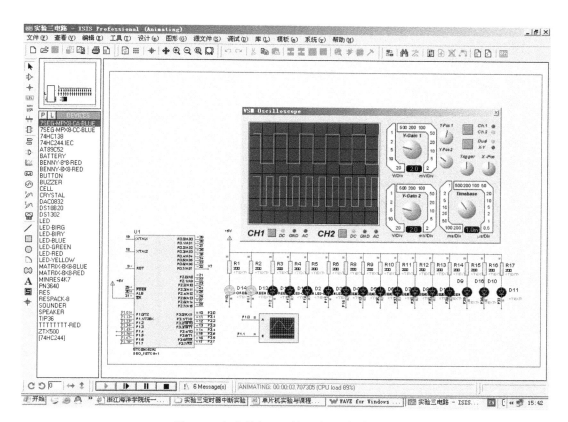

图 3.3 实验程序 3.2 的 Proteus 仿真效果

```
/*-----------------------------------------
   main function
----------------------------------------- */
void main(void)
{
TMOD = 0x01;
TH0 = 0xFC;
TL0 = 0x18;
TR0 = 1;
IE = 0x82;
while(1);
}


/*-----------------------------------------
   T0 interrupt function
----------------------------------------- */
```

```
void intt0(void) interrupt 1
{
TL0 = 0x18;
TH0 = 0xFC;
P10 = ! P10;
}


/*-------------------------------------
Timer interrupt program V3.2
MCU STC89C52RC   XAL 12MHz
Build by Gavin Hu, 2010.6.9
------------------------------------- */
# include <reg51.h>
sbit P10 = P1^0;
sbit P11 = P1^1;


/*-------------------------------------
  main function
------------------------------------- */
void main(void)
{
TMOD = 0x11;
TH0 = 0xFC;
TL0 = 0x18;
TH1 = 0xFE;
TL1 = 0x0C;
TR0 = 1;
TR1 = 1;
IE = 0x8A;
while(1);
}


/*-------------------------------------
  T0 interrupt function
------------------------------------- */
void intt0(void) interrupt 1
{
TL0 = 0x18;
TH0 = 0xFC;
```

```
P10 = ! P10;
}


/*--------------------------------------
   T1 interrupt function
--------------------------------------- */
void intt1(void) interrupt 3
{
TL1 = 0x0C;
TH1 = 0xFE;
P11 = ! P11;
}
```

第 4 章

实验四：串行口通信实验

4.1 实验内容与要求

1. 实验目的

(1)学习掌握串行口方式 0 及方式 1 工作模式下的程序初始化方法。

(2)学习掌握串行口数据发送及接收程序的编程方法。

2. Proteus 仿真实验硬件电路

串行口通信实验的 Proteus 仿真实验硬件电路如图 4.1 所示。

3. 实验任务

(1)在串行口方式 0 下将数据 1、2、3、4、5、6、7、8 依次从单片机串行口通过同步移位方式发送到串入/并出集成电路 74HC595，并在 74HC595 数据输出口用 LED 小灯显示数据(灯亮为逻辑 0，灯灭为逻辑 1)。

(2)在串行口方式 1(波特率为 1200B)下将数据 1、2、3、4、5、6、7、8 分别从一单片机发送到另一单片机，接收单片机在 P1 口输出接收到的数据，并用端口的 LED 小灯显示数据(灯亮为逻辑 0，灯灭为逻辑 1)，如图 4.2 所示。

4. 实验预习要求

(1)阅读教材中有关串行口基本结构及操作方式的内容，熟悉串行口在不同模式下的寄存器设置及程序的编写；查阅 74HC595 使用资料，掌握电路使用时的接法。

(2)计算波特率为 1200B 时的定时器初值。

(3)根据硬件电路原理图，画出实际接线图。

(4)掌握 Wave、Madwin、Keil-51 等编译软件的使用方法。

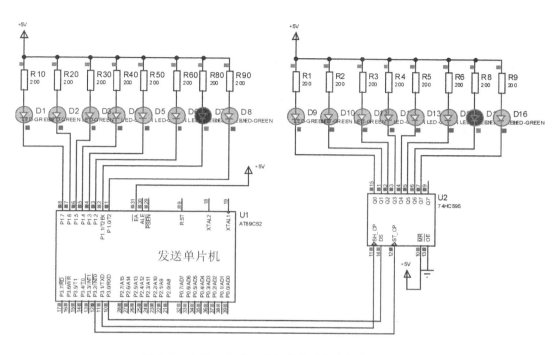

图 4.1 串行口方式 0 移位模式下实验仿真硬件电路

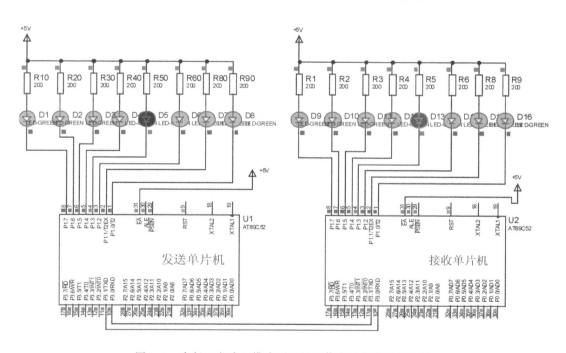

图 4.2 串行口方式 1 模式下双机通信实验仿真硬件电路

(5)根据实验任务设计出相应的调试程序。

(6)完成预习报告。

5. 实验设备

计算机(已安装单片机汇编编译软件及 Proteus 软件)。

6. 思考题

怎样用串行通信实现用一单片机控制另一单片机 P1 端口的 LED 小灯循环闪亮?

7. 实验报告要求

(1)整理好实验任务 1～2 中经 Proteus 运行正确的程序。

(2)解答思考题并进行程序的汇编与仿真调试,整理好经 Proteus 运行正确的思考题程序。

4.2 实验四参考汇编程序

```
;**************************************************;
;                 实验程序 4.1 发送程序              ;
;   在串行口方式 0 下将数据 1、2、3、4、5、6、7、8 依次从单片机串行口   ;
;   通过同步移位方式发送到串入/并出集成电路 74HC595,并在 74HC595   ;
;   数据输出口用 LED 小灯显示数据(灯亮为逻辑 0,灯灭为逻辑 1)   ;
;                    12MHz 晶振                    ;
;**************************************************;
;
        ORG     0000H           ;主程序执行入口地址
        LJMP    MAIN            ;跳至 MAIN 执行
;
MAIN:   MOV     30H,#01H        ;在 30H～37H 分别对应放数据 1～8
        MOV     31H,#02H        ;
        MOV     32H,#03H        ;
        MOV     33H,#04H        ;
        MOV     34H,#05H        ;
        MOV     35H,#06H        ;
        MOV     36H,#07H        ;
        MOV     37H,#08H        ;
        MOV     SCON,#00H       ;设串口为方式 0
```

```
              CLR     P3.2              ；74HC595 输出锁存
LOOP：        LCALL   UARTOUT           ；调用发送子程序
              AJMP    LOOP              ；
;
UARTOUT：MOV   R0,＃30H                 ；发送数据首址入 R0
         MOV   R2,＃8                   ；发送字节个数入 R2
SOUT：   MOV   A,@R0                    ；发送数据入 A
         MOV   P1,A                    ；放在 P1 口显示
         CLR   TI                      ；清发送标志 TI
         MOV   SBUF,A                  ；启动发送
WAITOUT：JNB   TI,WAITOUT              ；发送等待
         SETB  P3.2                    ；74HC595 将数据输出至端口 Q0～Q7
         CLR   P3.2                    ；
         LCALL DL1S                    ；延时 1s
         INC   R0                      ；指向下一字节
         DJNZ  R2,SOUT                 ；8 个字节未发完,转 SOUT
         RET                           ；8 个字节发完,结束
;
;延时程序(约 0.5ms)
DL503：   MOV    R7,＃250
LOOP1：   DJNZ   R7,LOOP1
          RET
;
;延时程序(约 10ms)
DL10ms：  MOV    R6,＃20
LOOP2：   LCALL  DL503
          DJNZ   R6,LOOP2
          RET
;
;延时程序(约 1s)
DL1S：    MOV    R5,＃100
LOOP3：   LCALL  DL10ms
          DJNZ   R5,LOOP3
          RET
          END                          ；结束
```

实验程序 4.1 的 Proteus 仿真效果图如图 4.3 所示。

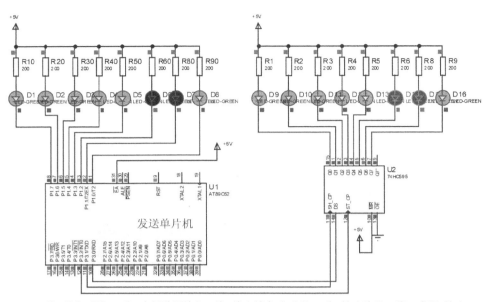

11脚—移位时钟, 14脚—串行数据输入,12脚—输出锁存(上升沿),10脚—输出清零,13脚—1高阻0输出

图 4.3 实验程序 4.1 的 Proteus 仿真效果

```
;**************************************************;
;              实验程序 4.2 发送程序              ;
;  在串行口方式 1 下将数据 1、2、3、4、5、6、7、8 分别从一单片机发送  ;
;  到另一单片机,接收单片机在 P1 口输出接收到的数据,并用端口的  ;
;  LED 小灯显示数据(灯亮为逻辑 0,灯灭为逻辑 1,波特率 1200)  ;
;                  12MHz 晶振                  ;
;**************************************************;
;
        ORG    0000H          ;主程序执行入口地址
        LJMP   MAIN           ;跳至 MAIN 执行
;
MAIN:   MOV    30H,♯01H       ;在 30H~37H 分别对应放数据 1~8
        MOV    31H,♯02H       ;
        MOV    32H,♯03H       ;
        MOV    33H,♯04H       ;
        MOV    34H,♯05H       ;
        MOV    35H,♯06H       ;
        MOV    36H,♯07H       ;
        MOV    37H,♯08H       ;
        MOV    TMOD,♯20H      ;T1 为 8 位自动重装模式
        MOV    TL1,♯0E6H      ;1200 波特率初值
        MOV    TH1,♯0E6H      ;
        CLR    ET1            ;关 T1 中断
```

```
            SETB    TR1                 ;开波特率发生器
LOOP：      LCALL   UARTOUT             ;调用发送子程序
            AJMP    LOOP                ;
;
UARTOUT：MOV        R0,#30H             ;发送数据首址入 R0
            MOV     R2,#8               ;发送字节个数入 R2
            MOV     SCON,#40H           ;设串口为方式 1
SOUT：      MOV     A,@R0               ;发送数据入 A
            MOV     P1,A                ;放在 P1 口显示
            CLR     TI                  ;清发送标志 TI
            MOV     SBUF,A              ;启动发送
WAITOUT：JNB       TI,WAITOUT           ;发送等待
            LCALL   DL1S                ;延时 1s
            INC     R0                  ;指向下一字节
            DJNZ    R2,SOUT             ;N 个字节未发完,转 SOUT
            RET                         ;N 个字节发完,结束
;
;延时程序(约 0.5ms)
DL503：     MOV     R7,#250
LOOP1：     DJNZ    R7,LOOP1
            RET
;
;延时程序(约 10ms)
DL10ms：    MOV     R6,#20
LOOP2：     LCALL   DL503
            DJNZ    R6,LOOP2
            RET
;
;延时程序(约 1s)
DL1S：      MOV     R5,#100
LOOP3：     LCALL   DL10ms
            DJNZ    R5,LOOP3
            RET
            END                         ;结束
```

```
;**************************************************;
;               实验程序 4.2 接收程序              ;
;  在串行口方式 1 下将数据 1、2、3、4、5、6、7、8 分别从一单片机发送  ;
;  到另一单片机,接收单片机在 P1 口输出接收到的数据,并用端口的  ;
;  LED 小灯显示数据(灯亮为逻辑 0,灯灭为逻辑 1,波特率为 1200)  ;
;                   12MHz 晶振                     ;
;**************************************************;
```

```
    ;
        ORG     0000H              ; 主程序执行入口地址
        LJMP    MAIN               ; 跳至 MAIN 执行
    ;
MAIN: MOV     TMOD,#20H           ; T1 为 8 位自动重装模式
        MOV     TL1,#0E6H          ; 1200 波特率初值
        MOV     TH1,#0E6H          ;
        MOV     SCON,#50H          ; 置串口为方式 1,REN = 1
        CLR     ET1                ; 关 T1 中断
        SETB    TR1                ; 开启波特率发生器
WAITIN:JNB     RI,WAITIN          ; 接收等待
        MOV     A,SBUF             ; 接收缓冲器数据入 A
        MOV     P1,A               ; 将接收数据放在 P1 口显示
        CLR     RI                 ; 清接收标志
        AJMP    WAITIN             ; 转接收等待
    ;
        END                        ; 结束
```

实验程序 4.2 的 Proteus 仿真效果如图 4.4 所示。

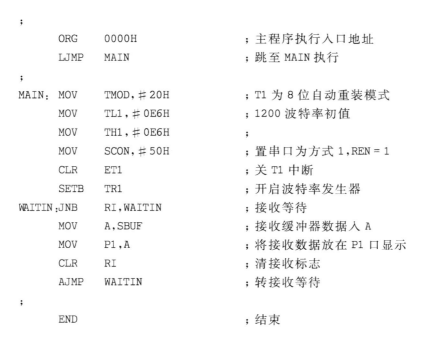
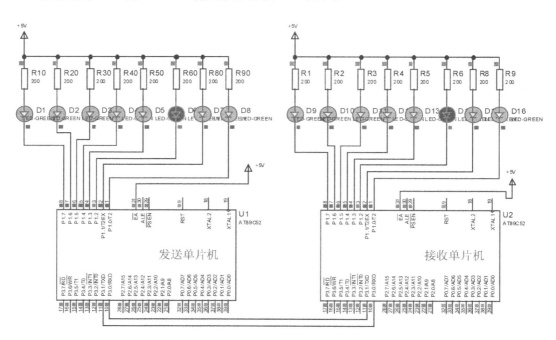

图 4.4　实验程序 4.2 的 Proteus 仿真效果

4.3　实验四参考 C 程序

```
/*------------------------------------
Communication program V4. 1
MCU STC89C52RC   XAL 12MHz
Build by Gavin Hu，2010. 6. 9
------------------------------------ */
# include ＜reg51. h＞
sbit HC595_ST = P3^2；
sbit HC595_SH = P3^1；
sbit HC595_DS = P3^0；
void delay_ms(unsigned int)；
/*------------------------------------
   main function
------------------------------------ */
void main(void)
{
unsigned char i；
SCON = 0x00；
HC595_ST = 0；
while(1)
    {
    for (i = 1；i＜ = 8；i ++ )
        {
        P1 = i；
        TI = 0；
        SBUF = i；
        while (! TI)；
        HC595_ST = 1；
        HC595_ST = 0；
        delay_ms(1000)；
        }
    }
}


/*------------------------------------
   Delay function
   Parameter：unsigned int dt
```

```
   Delay time = dt(ms)
---------------------------------------- */
void delay_ms(unsigned int dt)
{
register unsigned char bt,ct;
for (;dt;dt--)
    for (ct = 2;ct;ct--)
        for (bt = 250; -- bt;);
}


/*----------------------------------------
Communication program V4.2 - R
MCU STC89C52RC   XAL 12MHz
Build by Gavin Hu, 2010.6.9
---------------------------------------- */
# include <reg51.h>


/*----------------------------------------
   main function
---------------------------------------- */
void main(void)
{
TMOD = 0x20;
TH1 = 0xE6;
TR1 = 1;
SCON = 0x50;
while(1)
    {
    while (!RI);
    RI = 0;
    P1 = SBUF;
    }
}


/*----------------------------------------
Communication program V4.2 - T
MCU STC89C52RC   XAL 12MHz
Build by Gavin Hu, 2010.6.9
---------------------------------------- */
# include <reg51.h>
void delay_ms(unsigned int);
```

```
/*----------------------------------------
   main function
   ---------------------------------------- */
void main(void)
{
unsigned char i;
TMOD = 0x20;
TH1 = 0xE6;
TR1 = 1;
SCON = 0x40;
while(1)
    {
    for (i = 1; i <= 8; i ++ )
        {
        P1 = i;
        TI = 0;
        SBUF = i;
        while (! TI);
        delay_ms(1000);
        }
    }
}
```

第 5 章

实验五:按键接口实验

5.1 实验内容与要求

1. 实验目的

(1)熟悉单片机简单按键及行列式按键的接口方法。

(2)掌握按键扫描及处理程序的编程方法和调试方法。

2. Proteus 仿真实验硬件电路

按键接口实验的 Proteus 仿真实验硬件电路如图 5.1 所示。

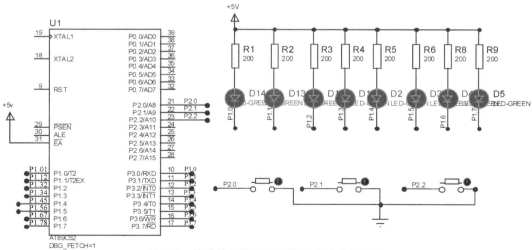

图 5.1 简单端口按键查询实验仿真硬件电路

3. 实验任务

(1)用单片机 P2.0～P2.2 端口的 3 个按键分别对应控制 P1.0～P1.3 端口的 3 个 LED 小灯的亮与灭。

(2)用单片机 P2.0～P2.7 端口的 16 个行列式按键分别对应控制 P1 与 P3 端口的 16 个 LED 小灯的亮与灭。

4. 实验预习要求

(1)根据硬件电路原理图,画出实际接线图。

(2)阅读教材中有关简单按键查询及行列式按键查询程序,了解程序设计方法;写出 4×4 行列式查询时用的与键号对应的键值表;根据实验任务设计相应的调试程序。 行列式按键查询实验的仿真硬件电路如图 5.2 所示。

(3)阅读掌握 Wave、Madwin、Keil-51 等编译软件的使用方法。

(4)完成预习报告。

5. 实验设备

计算机(已安装单片机汇编编译软件及 Proteus 软件)。

6. 实验报告要求

整理好实验任务 1～2 中经 Proteus 运行正确的程序。

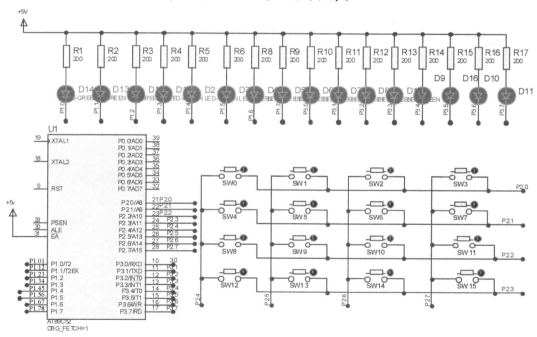

图 5.2 行列式按键查询实验仿真硬件电路

5.2　实验五参考汇编程序

```
;**********************************************************;
;                      实验程序 5.1                       ;
;    用单片机 P2.0～P2.2 端口的 3 个按键分别控制 P1.0～P1.3   ;
;    端口的 3 个 LED 小灯的亮与灭                           ;
;                      12MHz 晶振                          ;
;**********************************************************;
;
KEYSW0    EQU    P2.0      ;按键 0
KEYSW1    EQU    P2.1      ;按键 1
KEYSW2    EQU    P2.2      ;按键 2
LED0      EQU    P1.0      ;LED 小灯 0
LED1      EQU    P1.1      ;LED 小灯 1
LED2      EQU    P1.2      ;LED 小灯 2
;
;***********;
; 主程序入口 ;
;***********;
;
ORG      0000H      ;程序执行开始地址
LJMP     START      ;跳至 START 执行
;
;***********;
;  主 程 序  ;
;***********;
;
START: MOV     P2,♯0FFH        ;置 P2 口为输入状态
KLOOP: JNB     KEYSW0,KEY0      ;读 KEYSW0 口,若为 0 转 KEY0
       JNB     KEYSW1,KEY1      ;读 KEYSW1 口,若为 0 转 KEY1
       JNB     KEYSW2,KEY2      ;读 KEYSW2 口,若为 0 转 KEY2
       AJMP    KLOOP            ;子程序返回
;
;0 键处理程序
KEY0:  LCALL   DL10ms           ;延时 10ms 消抖
       JB      KEYSW0,KLOOP     ;KEYSW0 为 1,程序返回(干扰)
```

```
            CPL      LED0              ;开 LED0 灯
WAIT0：JNB      KEYSW0,WAIT0       ;等待键释放
            LCALL    DL10ms            ;延时消抖
            JNB      KEYSW0,WAIT0      ;
            AJMP     KLOOP             ;返回主程序
      ;
;1 键处理程序
KEY1：LCALL    DL10ms            ;延时 10ms 消抖
            JB       KEYSW1,KLOOP      ;KEYSW1 为 1,程序返回(干扰)
            CPL      LED1              ;开 LED1 灯
WAIT1：JNB      KEYSW1,WAIT1      ;等待键释放
            LCALL    DL10ms            ;延时消抖
            JNB      KEYSW1,WAIT1      ;
            AJMP     KLOOP             ;返回主程序
      ;
;2 键处理程序
KEY2：LCALL    DL10ms            ;延时 10ms 消抖
            JB       KEYSW2,KLOOP      ;KEYSW2 为 1,程序返回(干扰)
            CPL      LED2              ;开 LED2 灯
WAIT2：JNB      KEYSW2,WAIT2      ;等待键释放
            LCALL    DL10ms            ;延时消抖
            JNB      KEYSW2,WAIT2      ;
            AJMP     KLOOP             ;返回主程序
      ;
;************;
;   延时程序   ;
;************;
;约 0.5ms 延时子程序,执行一次时间为 513 μs
DL512：MOV      R2,#0FFH
LOOP1：DJNZ     R2,LOOP1
            RET
      ;
;约 10ms 延时子程序(调用 20 次 0.5ms 延时子程序)
DL10ms:MOV      R3,#14H
LOOP2：LCALL    DL512
            DJNZ     R3,LOOP2
            RET


            END                       ;程序结束
```

实验程序 5.1 的 Proteus 仿真效果图如图 5.3 所示。

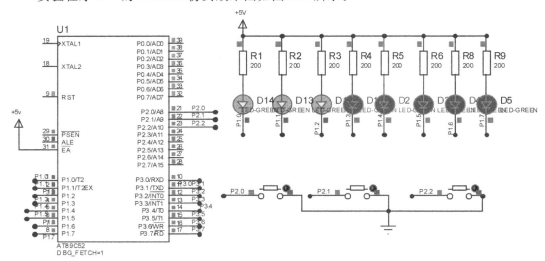

图 5.3   实验程序 5.1 的 Proteus 仿真效果

```
;**********************************************************;
;                     实验程序 5.2                        ;
;   用单片机 P2.0～P2.7 端口的 16 个行列式按键分别控制 P1 与 P3 ;
;   端口的 16 个 LED 小灯的亮与灭                            ;
;                     12MHz 晶振                           ;
;**********************************************************;
;
;按键口及小灯口的定义
          KEY     EQU     P2       ;定义 P2 口为行列式按键口
          LED0    EQU     P1.0     ;定义小灯名
          LED1    EQU     P1.1     ;定义小灯名
          LED2    EQU     P1.2     ;定义小灯名
          LED3    EQU     P1.3     ;定义小灯名
          LED4    EQU     P1.4     ;定义小灯名
          LED5    EQU     P1.5     ;定义小灯名
          LED6    EQU     P1.6     ;定义小灯名
          LED7    EQU     P1.7     ;定义小灯名
          LED8    EQU     P3.0     ;定义小灯名
          LED9    EQU     P3.1     ;定义小灯名
          LED10   EQU     P3.2     ;定义小灯名
          LED11   EQU     P3.3     ;定义小灯名
          LED12   EQU     P3.4     ;定义小灯名
          LED13   EQU     P3.5     ;定义小灯名
          LED14   EQU     P3.6     ;定义小灯名
          LED15   EQU     P3.7     ;定义小灯名
```

```
                KEYWORD EQU     23H      ;键值寄放单元
;程序入口地址定位
                ORG     0000H                    ;程序开始地址
                LJMP    MAIN                     ;转 MAIN 执行
;
;主程序
MAIN：          LCALL   KEYWORK                  ;调查键子程序
                AJMP    MAIN                     ;转 MAIN 循环
;
;4×4 行列式扫描查键及功能子程序
KEYWORK：       MOV     KEY,♯0FFH                ;置 KEY 口为输入状态
                CLR     KEY.0                    ;扫描第一行(第一行为 0)
                MOV     A,KEY                    ;读入 KEY 口值
                ANL     A,♯0F0H                  ;低四位为 0
                CJNE    A,♯0F0H,KEYCON           ;有键按下转 KEYCON
                SETB    KEY.0                    ;扫描第二行(第二行为 0)
                CLR     KEY.1                    ;
                MOV     A,KEY                    ;读入 KEY 口值
                ANL     A,♯0F0H                  ;低四位为 0
                CJNE    A,♯0F0H,KEYCON           ;有键按下转 KEYCON
                SETB    KEY.1                    ;扫描第三行(第三行为 0)
                CLR     KEY.2                    ;
                MOV     A,KEY                    ;读入 KEY 口值
                ANL     A,♯0F0H                  ;低四位为 0
                CJNE    A,♯0F0H,KEYCON           ;有键按下转 KEYCON
                SETB    KEY.2                    ;扫描第四行(第四行为 0)
                CLR     KEY.3                    ;
                MOV     A,KEY                    ;读入 KEY 口值
                ANL     A,♯0F0H                  ;低四位为 0
                CJNE    A,♯0F0H,KEYCON           ;有键按下转 KEYCON
                SETB    KEY.3                    ;结束行扫描
                RET                              ;子程序返回
KEYCON：        LCALL   DL10ms                   ;消抖处理
                MOV     A,KEY                    ;再读入 KEY 口值
                ANL     A,♯0F0H                  ;低四位为 0
                CJNE    A,♯0F0H,KEYCHE           ;确认键按下,转 KEYCHE
KEYOUT：        RET                              ;干扰,子程序返回
KEYCHE：        MOV     A,KEY                    ;读 KEY 口值
                MOV     KEYWORD,A                ;键值暂存
CJLOOP：
                MOV     A,KEY                    ;读 KEY 口值
```

```
                    ANL     A,♯0F0H                    ;低四位为 0
                    CJNE    A,♯0F0H,CJLOOP             ;键还按着,转 CJLOOP 等待释放
                    MOV     R7,♯00H                    ;键释放,置 R7 初值为♯00H(查表次数)
                    MOV     DPTR,♯KEYTAB               ;取键值表首址
        CHEKEYLOOP: MOV     A,R7                       ;查表次数入 A
                    MOVC    A,@A+DPTR                  ;查表
                    XRL     A,KEYWORD                  ;查表值与 KEY 口读入值比较
                    JZ      KEYOK                      ;为 0(相等)转 KEYOK
                    INC     R7                         ;不等,查表次数加 1
                    CJNE    R7,♯10H,CHEKEYLOOP         ;查表次数不超过 16 次,转 CHEKEYLOOP 再查
                    RET                                ;16 次到,退出
        ;
        KEYOK:      MOV     A,R7                       ;查表次数入 A(即键号值)
                    MOV     B,A                        ;放入 B
                    RL      A                          ;左移
                    ADD     A,B                        ;相加(键号乘 3 处理 JMP 3 字节指令)
                    MOV     DPTR,♯KEYFUNTAB            ;取键功能散转表首址
                    JMP     @A+DPTR                    ;查表
        KEYFUNTAB:  LJMP    KEYFUN00                   ;键功能散转表,跳至 0 号键功能程序
                    LJMP    KEYFUN01                   ;跳至 01 号键功能程序
                    LJMP    KEYFUN02                   ;跳至 02 号键功能程序
                    LJMP    KEYFUN03
                    LJMP    KEYFUN04
                    LJMP    KEYFUN05
                    LJMP    KEYFUN06
                    LJMP    KEYFUN07
                    LJMP    KEYFUN08
                    LJMP    KEYFUN09
                    LJMP    KEYFUN10
                    LJMP    KEYFUN11
                    LJMP    KEYFUN12
                    LJMP    KEYFUN13
                    LJMP    KEYFUN14
                    LJMP    KEYFUN15                   ;跳至 15 号键功能程序
                    RET                                ;散转出错返回
        ;
        ;键号对应 KEY 口数值表(同时按下两键为无效操作)
        KEYTAB:     DB      0EEH,0DEH,0BEH,7EH,0EDH,0DDH,0BDH,7DH
                    DB      0EBH,0DBH,0BBH,7BH,0E7H,0D7H,0B7H,77H,0FFH,0FFH
        ;
        ;00 号键功能程序
```

```
KEYFUN00：      CPL LED0          ;开关小灯
                RET               ;返回
;01 号键功能程序
KEYFUN01：      CPL LED1          ;开关小灯
                RET               ;返回
;02 号键功能程序
KEYFUN02：      CPL LED2          ;开关小灯
                RET               ;返回
;03 号键功能程序
KEYFUN03：      CPL LED3          ;开关小灯
                RET               ;返回
;04 号键功能程序
KEYFUN04：      CPL LED4          ;开关小灯
                RET               ;返回
;05 号键功能程序
KEYFUN05：      CPL LED5          ;开关小灯
                RET               ;返回
;06 号键功能程序
KEYFUN06：      CPL LED6          ;开关小灯
                RET               ;返回
;07 号键功能程序
KEYFUN07：      CPL LED7          ;开关小灯
                RET               ;返回
;08 号键功能程序
KEYFUN08：      CPL LED8          ;开关小灯
                RET               ;返回
;09 号键功能程序
KEYFUN09：      CPL LED9          ;开关小灯
                RET               ;返回
;10 号键功能程序
KEYFUN10：      CPL LED10         ;开关小灯
                RET               ;返回
;11 号键功能程序
KEYFUN11：      CPL LED11         ;开关小灯
                RET               ;返回
;12 号键功能程序
KEYFUN12：      CPL LED12         ;开关小灯
                RET               ;返回
;13 号键功能程序
KEYFUN13：      CPL LED13         ;开关小灯
                RET               ;返回
```

;14 号键功能程序

| KEYFUN14: | CPL LED14 | ;开关小灯 |
| | RET | ;返回 |

;15 号键功能程序

| KEYFUN15: | CPL LED15 | ;开关小灯 |
| | RET | ;返回 |

;

;513 μs延时子程序

DL513:	MOV	R3,♯0FFH
DL513LOOP:	DJNZ	R3,DL513LOOP
	RET	

;

;10ms 延时子程序(消抖动用)

DL10ms:	MOV	R6,♯20
DL10msLOOP:	LCALL DL513	
	DJNZ	R6,DL10msLOOP
	RET	

;

| | END | ;程序结束 |

实验程序 5.2 的 Proteus 仿真效果如图 5.4 所示。

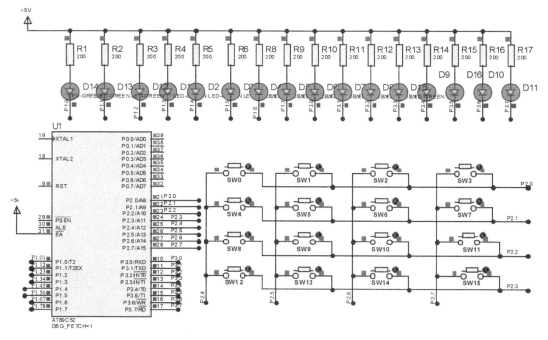

图 5.4　实验程序 5.2 的 Proteus 仿真效果

5.3　实验五参考 C 程序

```
/*------------------------------------
Key program V5.1
MCU STC89C52RC   XAL 12MHz
Build by Gavin Hu，2010.6.9
------------------------------------ */
# include <reg51.h>
sbit KEY1 = P2^0;
sbit KEY2 = P2^1;
sbit KEY3 = P2^2;
sbit LED1 = P1^0;
sbit LED2 = P1^1;
sbit LED3 = P1^2;
void delay_ms(unsigned int);

/*------------------------------------
  main function
------------------------------------ */
void main(void)
{
while(1)
    {
    if (KEY1 == 0)
        {
        LED1 = ! LED1;
        delay_ms(500);
        }
    if (KEY2 == 0)
        {
        LED2 = ! LED2;
        delay_ms(500);
        }
    if (KEY3 == 0)
        {
        LED3 = ! LED3;
```

```
            delay_ms(500);
            }
        }
}


/*------------------------------------
   Delay function
   Parameter: unsigned int dt
   Delay time = dt(ms)
   ------------------------------------ */
void delay_ms(unsigned int dt)
{
register unsigned char bt,ct;
for (;dt;dt --)
    for (ct = 2;ct;ct --)
      for (bt = 250; -- bt;);
}


/*------------------------------------
Key program V5. 2
MCU STC89C52RC   XAL 12MHz
Build by Gavin Hu, 2010. 6. 9
------------------------------------ */
# include <reg51. h>
sbit LED1 = P1^0;
sbit LED2 = P1^1;
sbit LED3 = P1^2;
sbit LED4 = P1^3;
sbit LED5 = P1^4;
sbit LED6 = P1^5;
sbit LED7 = P1^6;
sbit LED8 = P1^7;
sbit LED9 = P3^0;
sbit LED10 = P3^1;
sbit LED11 = P3^2;
sbit LED12 = P3^3;
sbit LED13 = P3^4;
sbit LED14 = P3^5;
```

```
sbit LED15 = P3^6;
sbit LED16 = P3^7;
sbit P24 = P2^4;
sbit P25 = P2^5;
sbit P26 = P2^6;
sbit P27 = P2^7;
void delay_ms(unsigned int);

/*--------------------------------------
   main function
-------------------------------------- */
void main(void)
{
while(1)
    {
    P2 = 0xfe;
    if (P24 == 0) {LED1 = ! LED1;delay_ms(500);}
        else if (P25 == 0) {LED2 = ! LED2;delay_ms(500);}
        else if (P26 == 0) {LED3 = ! LED3;delay_ms(500);}
        else if (P27 == 0) {LED4 = ! LED4;delay_ms(500);}
    P2 = 0xfd;
    if (P24 == 0) {LED5 = ! LED5;delay_ms(500);}
        else if (P25 == 0) {LED6 = ! LED6;delay_ms(500);}
        else if (P26 == 0) {LED7 = ! LED7;delay_ms(500);}
        else if (P27 == 0) {LED8 = ! LED8;delay_ms(500);}
    P2 = 0xfb;
    if (P24 == 0) {LED9 = ! LED9;delay_ms(500);}
        else if (P25 == 0) {LED10 = ! LED10;delay_ms(500);}
        else if (P26 == 0) {LED11 = ! LED11;delay_ms(500);}
        else if (P27 == 0) {LED12 = ! LED12;delay_ms(500);}
    P2 = 0xf7;
    if (P24 == 0) {LED13 = ! LED13;delay_ms(500);}
        else if (P25 == 0) {LED14 = ! LED14;delay_ms(500);}
        else if (P26 == 0) {LED15 = ! LED15;delay_ms(500);}
        else if (P27 == 0) {LED16 = ! LED16;delay_ms(500);}
    }
}
```

```
/*-------------------------------------

   Delay function

   Parameter: unsigned int dt

   Delay time = dt(ms)

-------------------------------------- */

void delay_ms(unsigned int dt)

{

register unsigned char bt,ct;

for (;dt;dt --)

    for (ct = 2;ct;ct --)

        for (bt = 250; -- bt;);

}
```

第 6 章

实验六：八位共阳 LED 数码管实验

6.1 实验内容与要求

1. 实验目的

(1)学习用动态扫描法实现八位 LED 共阳数码管的数字显示。

(2)学习掌握七段共阳数码管的小数点显示方法。

2. Proteus 仿真实验硬件电路

八位共阳 LED 数据管实验的仿真硬件电路如图 6.1 所示。

3. 实验任务

(1)将 8 个内存单元中的数(1~8)用 8 个 LED 共阳数码管显示出来。

(2)在显示数的百位及万位位置显示 2 个小数点。

(3)让某个内存中的数闪烁显示。

4. 实验预习要求

(1)学习掌握 LED 段码显示器动态扫描显示原理；了解七段 LED 显示器中小数点的显示方法；分析显示 0~9 数字及个别英文字母(如 A、C、E、F、H 等)时对应的显示段码数据。

(2)根据硬件电路原理图,画出实际仿真接线图。

(3)根据实验任务设计出相应的调试程序。

(4)完成预习报告。

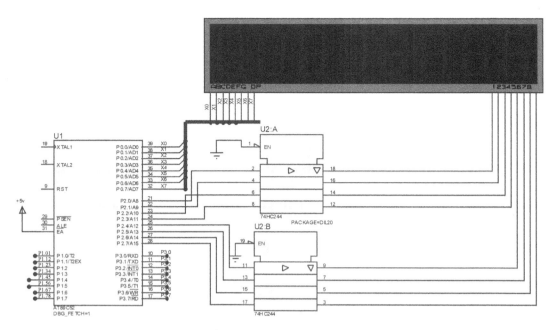

图 6.1　八位共阳 LED 数码管实验仿真硬件电路

5. 实验设备

计算机(已安装单片机汇编编译软件及 Proteus 软件)。

6. 思考题

让某个内存中的数闪烁显示,除了你使用的方法外还有什么方法,各有什么优缺点?

7. 实验报告要求

(1)整理好实验任务 1~3 中经 Proteus 运行正确的程序。
(2)解答思考题。

6.2　实验六参考汇编程序

```
;************************************************;
;                     实验程序 6.1                ;
;     将 8 个内存单元中的数(1~8)用 8 个 LED 共阳数码管显示出来    ;
;                     12MHz 晶振                  ;
;************************************************;
;显示首地址定义
```

```
            DISPFIRST    EQU   30H        ;显示首址存放单元
;
;主程序入口地址定义
            ORG          0000H        ;程序执行开始地址
            LJMP         START        ;跳到标号 START 执行
;
;以下为主程序开始
;
    START： MOV     DISPFIRST,#70H      ;显示单元为 70H~77H
            MOV     70H,#8
            MOV     71H,#7
            MOV     72H,#6
            MOV     73H,#5
            MOV     74H,#4
            MOV     75H,#3
            MOV     76H,#2
            MOV     77H,#1
;以下为主程序循环
    START1： LCALL   DISPLAY             ;调用显示子程序
            AJMP    START1             ;
;
;***************************************;
;          八位共阳 LED 显示程序          ;
;***************************************;
;显示数据在 70H~77H 单元内,用八位 LED 共阳数码管显示,P0 口输出
;段码数据,P2 口作扫描控制,每个 LED 数码管亮 1ms 时间再逐位循环
    DISPLAY：MOV   R1,DISPFIRST         ;指向显示数据首址
            MOV   R5,#80H             ;扫描控制字初值
    PLAY：  MOV   A,R5               ;扫描字放入 A
            MOV   P2,A               ;从 P2 口输出
            MOV   A,@R1              ;取显示数据到 A
            MOV   DPTR,#TAB          ;取段码表地址
            MOVC  A,@A+DPTR          ;查显示数据对应段码
            MOV   P0,A               ;段码放入 P0 口
            LCALL DL1ms              ;显示 1ms
            INC   R1                 ;指向下一地址
            MOV   A,R5               ;扫描控制字放入 A
            JB    ACC.0,ENDOUT       ;ACC.0 = 1 时一次显示结束
            RR    A                  ;A 中数据循环右移
```

```
                MOV      R5,A              ;放回 R5 内
                MOV      P0,♯0FFH
                AJMP     PLAY              ;跳回 PLAY 循环
        ENDOUT: MOV      P2,♯00H           ;一次显示结束,P2 口复位
                MOV      P0,♯0FFH          ;P0 口复位
                RET                        ;子程序返回
TAB: DB 0C0H,0F9H,0A4H,0B0H,99H,92H,82H,0F8H,80H,90H,0FFH,88H,0BFH
;共阳段码表 "0""1""2" "3""4""5""6""7" "8""9""不亮""A""-"
;***************************************;
;                 延时程序                 ;
;***************************************;
;
;1ms 延时程序,LED 显示程序
        DL1ms:  MOV      R6,♯14H
        DL1:    MOV      R7,♯19H
        DL2:    DJNZ     R7,DL2
                DJNZ     R6,DL1
                RET
;
        END          ;程序结束
;***************************************************************
```

实验程序 6.1 的 Proteus 仿真效果如图 6.2 所示。

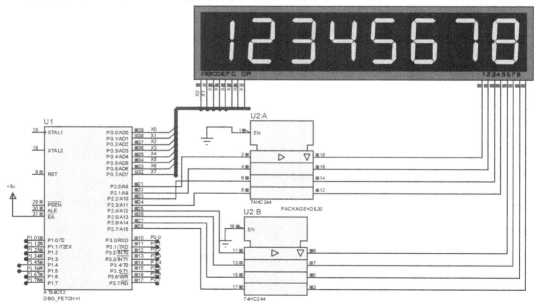

图 6.2   实验程序 6.1 的 Proteus 仿真效果

```
;**********************************************************;
;                     实验程序6.2                          ;
;     将8个内存单元中的数(2~9)用8个LED共阳数码管显示出来     ;
;     在显示数的百位及万位位置显示2个小数点                  ;
;                     12MHz晶振                            ;
;**********************************************************;
;显示首地址定义
        DISPFIRST    EQU   30H          ;显示首址存放单元
;
;主程序入口地址定义
        ORG          0000H              ;程序执行开始地址
        LJMP         START              ;跳到标号START执行
;
;以下为主程序开始
;
        START:  MOV  DISPFIRST,#70H     ;显示单元为70H~77H
                MOV  70H,#2
                MOV  71H,#3
                MOV  72H,#4
                MOV  73H,#5
                MOV  74H,#6
                MOV  75H,#7
                MOV  76H,#8
                MOV  77H,#9
;以下为主程序循环
        START1: LCALL  DISPLAY          ;调用显示子程序
                AJMP   START1           ;
;
;**********************************************;
;          八位共阳LED显示程序                 ;
;**********************************************;
;显示数据在70H~77H单元内,用八位LED共阳数码管显示,P0口
;输出段码数据,P2口作扫描控制,每个LED数码管亮1ms时间再逐位循环
        DISPLAY:MOV  R1,DISPFIRST       ;指向显示数据首址
                MOV  R5,#80H            ;扫描控制字初值
        PLAY:   MOV  A,R5              ;扫描字放入A
                MOV  P2,A              ;从P2口输出
                MOV  A,@R1             ;取显示数据到A
                MOV  DPTR,#TAB         ;取段码表地址
```

```
            MOVC    A,@A + DPTR          ; 查显示数据对应段码
            MOV     P0,A                 ; 段码放入 P0 口
            MOV     A,R5                 ;
            JNB     ACC.5,LOOP5          ; 小数点处理
            CLR     P0.7                 ;
    LOOP5:  JNB     ACC.3,LOOP6          ; 小数点处理
            CLR     P0.7                 ;
    LOOP6:  LCALL   DL1ms                ; 显示 1ms
            INC     R1                   ; 指向下一地址
            MOV     A,R5                 ; 扫描控制字放入 A
            JB      ACC.0,ENDOUT         ; ACC.0 = 1 时一次显示结束
            RR      A                    ; A 中数据循环右移
            MOV     R5,A                 ; 放回 R5 内
            MOV     P0,♯0FFH
            AJMP    PLAY                 ; 跳回 PLAY 循环
    ENDOUT: MOV     P2,♯00H              ; 一次显示结束,P2 口复位
            MOV     P0,♯0FFH             ; P0 口复位
            RET                          ; 子程序返回
TAB: DB 0C0H,0F9H,0A4H,0B0H,99H,92H,82H,0F8H,80H,90H,0FFH,88H,0BFH
;共阳段码表"0""1""2" "3""4""5""6""7" "8""9""不亮""A"" - "
;
;***********************************;
;            延时程序                 ;
;***********************************;
;
;1ms 延时程序,LED 显示程序用
            DL1ms:  MOV     R6,♯14H
            DL1:    MOV     R7,♯19H
            DL2:    DJNZ    R7,DL2
                    DJNZ    R6,DL1
                    RET
            END             ;程序结束
;*******************************************************************
```

实验程序 6.2 的 Proteus 仿真效果如图 6.3 所示。

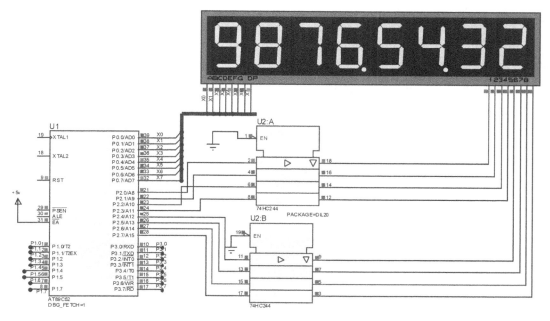

图 6.3　实验程序 6.2 的 Proteus 仿真效果

```
;**************************************************;
;                    实验程序6.3                  ;
;    将8个内存单元中的数(2～9)用8个LED共阳数码管显示出来  ;
;    在显示数的百位及万位位置显示2个小数点,让某个内存中的  ;
;    数闪烁显示                                    ;
;                   12MHz 晶振                     ;
;**************************************************;
;显示首地址定义
        DISPFIRST    EQU  30H        ;显示首址存放单元
;
;主程序入口地址定义
        ORG          0000H           ;程序执行开始地址
        LJMP         START           ;跳到标号START执行
;
;以下为主程序开始
;
        START:  MOV  DISPFIRST,#70H  ;显示单元为70H～77H
                MOV  70H,#2
                MOV  71H,#3
                MOV  72H,#4
                MOV  73H,#5
```

```
        MOV     74H,#6
        MOV     75H,#7
        MOV     76H,#8
        MOV     77H,#9
        MOV     78H,#10              ; 内存为 10 时显示段码为"不显示"
; 以下为主程序循环
START1: MOV     R4,#50               ; 闪烁间隔控制:50×8ms = 400ms
DISLOOP:LCALL   DISPLAY              ; 调用显示子程序
        DJNZ    R4,DISLOOP           ;
        XCH     A,78H                ; 以下 78H 与 76H 中数进行交换
        XCH     A,76H                ; 76H 单元的数闪烁
        XCH     A,78H                ;
        AJMP    START1               ;
;
;*******************************************;
;          八位共阳 LED 显示程序            ;
;*******************************************;
; 显示数据在 70H~77H 单元内,用八位 LED 共阳数码管显示,P0 口输出段码数据,P2 口作
; 扫描控制,每个 LED 数码管亮 1ms 时间再逐位循环
DISPLAY:MOV     R1,DISPFIRST         ; 指向显示数据首址
        MOV     R5,#80H              ; 扫描控制字初值
PLAY:   MOV     A,R5                 ; 扫描字放入 A
        MOV     P2,A                 ; 从 P2 口输出
        MOV     A,@R1                ; 取显示数据到 A
        MOV     DPTR,#TAB            ; 取段码表地址
        MOVC    A,@A+DPTR            ; 查显示数据对应段码
        MOV     P0,A                 ; 段码放入 P0 口
        MOV     A,R5                 ;
        JNB     ACC.5,LOOP5          ; 小数点处理
        CLR     P0.7                 ;
LOOP5:  JNB     ACC.3,LOOP6          ; 小数点处理
        CLR     P0.7                 ;
LOOP6:  LCALL   DL1ms                ; 显示 1ms
        INC     R1                   ; 指向下一地址
        MOV     A,R5                 ; 扫描控制字放入 A
        JB      ACC.0,ENDOUT         ; ACC.0 = 1 时一次显示结束
        RR      A                    ; A 中数据循环右移
        MOV     R5,A                 ; 放回 R5 内
        MOV     P0,#0FFH             ;
        AJMP    PLAY                 ; 跳回 PLAY 循环
```

```
ENDOUT: MOV     P2,♯00H          ;一次显示结束,P2 口复位
        MOV     P0,♯0FFH         ;P0 口复位
        RET                      ;子程序返回
TAB: DB 0C0H,0F9H,0A4H,0B0H,99H,92H,82H,0F8H,80H,90H,0FFH,88H,0BFH
;共阳段码表"0""1""2" "3""4""5""6""7" "8""9""不亮""A"" - "
;
;**********************************;
;              延时程序             ;
;**********************************;
;1ms 延时程序,LED 显示程序用
        DL1ms: MOV    R6,♯14H
        DL1：  MOV    R7,♯19H
        DL2：  DJNZ   R7,DL2
               DJNZ   R6,DL1
               RET
;
               END    ;程序结束
;****************************************************************
```

实验程序 6.3 的 Proteus 仿真效果如图 6.4 所示。

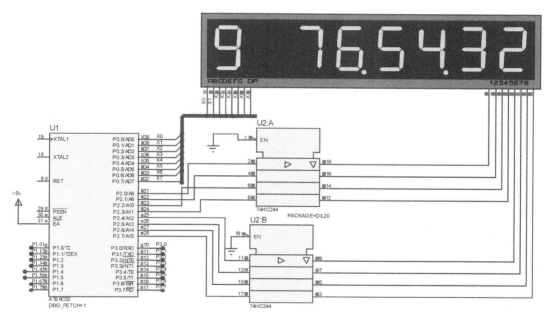

图 6.4　实验程序 6.3 的 Proteus 仿真效果

6.3　实验六参考 C 程序

```
/*----------------------------------------
Digit LED display program V6.1
MCU STC89C52RC  XAL 12MHz
Build by Gavin Hu, 2010.6.9
---------------------------------------- */
# include <reg51.h>
void delay_ms(unsigned int);
void display(char * );


/*----------------------------------------
   main function
---------------------------------------- */
void main(void)
{
char display_ram[] = {1,2,3,4,5,6,7,8};
while(1)
    {
    display(display_ram);
    }
}


/*----------------------------------------
   Display function
   8 digit LED tubes
   Parameter: sting point to display
---------------------------------------- */
void display(char * disp_ram)
{
unsigned char i;
unsigned char code table[] =
{0xc0,0xf9,0xa4,0xb0,0x99,0x92,0x82,0xf8,0x80,0x90,0x88,
0x83,0xc6,0xa1,0x86,0x8e,0xbf,0xff};
for (i = 0;i<8;i ++ )
    {
    P0 = table[disp_ram[i]];
```

```c
    P2 = 0x01<<i;
    delay_ms(1);
    P0 = 0xff;
    P2 = 0;
    }
}


/*--------------------------------
  Delay function
  Parameter: unsigned int dt
  Delay time = dt(ms)
----------------------------------*/
void delay_ms(unsigned int dt)
{
register unsigned char bt,ct;
for (;dt;dt--)
    for (ct=2;ct;ct--)
        for (bt=250;--bt;);
}
/*--------------------------------
Digit LED display program V6.2
MCU STC89C52RC   XAL 12MHz
Build by Gavin Hu, 2010.6.9
----------------------------------*/
#include <reg51.h>
void delay_ms(unsigned int);
void display(char*);


/*--------------------------------
  main function
----------------------------------*/
void main(void)
{
char display_ram[] = {1,2,3,4,5,6,7,8};
while(1)
    {
    display(display_ram);
    }
}
```

```c
/*----------------------------------------
   Display function
   8 digit LED tubes
   Parameter: sting point to display
   ------------------------------------ */
void display(char * disp_ram)
{
unsigned char i;
unsigned char code table[] =
{0xc0,0xf9,0xa4,0xb0,0x99,0x92,0x82,0xf8,0x80,0x90,0x88,
0x83,0xc6,0xa1,0x86,0x8e,0xbf,0xff};
for (i = 0;i<8;i++ )
    {
    P0 = table[disp_ram[i]];
    if ((i == 3)||(i == 5)) P0 & = 0x7F;
    P2 = 0x01<<i;
    delay_ms(1);
    P0 = 0xff;
    P2 = 0;
    }
}


/*----------------------------------------
   Delay function
   Parameter: unsigned int dt
   Delay time = dt(ms)
   ------------------------------------ */
void delay_ms(unsigned int dt)
{
register unsigned char bt,ct;
for (;dt;dt--)
    for (ct = 2;ct;ct--)
        for (bt = 250; -- bt;);
}


/*----------------------------------------
Digit LED display program V6.3
MCU STC89C52RC   XAL 12MHz
Build by Gavin Hu, 2010.6.9
   ------------------------------------ */
```

```
#include <reg51.h>
void delay_ms(unsigned int);
void display(char *);


/*----------------------------------
   main function
----------------------------------- */
void main(void)
{
char display_ram[] = {1,2,3,4,5,6,7,8};
display_ram[7] |= 0x80;
while(1)
    {
    display(display_ram);
    }
}
/*----------------------------------
   Display function
   8 digit LED tubes
   Parameter: sting point to display
   Bit-7 of display data set means Twinkling
---------------------------------- */
void display(char * disp_ram)
{
static unsigned char disp_count;
unsigned char i;
unsigned char code table[] =
{0xc0,0xf9,0xa4,0xb0,0x99,0x92,0x82,0xf8,0x80,0x90,0x88,
0x83,0xc6,0xa1,0x86,0x8e,0xbf,0xff};
disp_count = (disp_count + 1)&0x7f;
for (i = 0;i<8;i++)
    {
    if (disp_ram[i]&0x80) P0 = (disp_count>32)? table[disp_ram[i]&0x7f]:0xff;
            else P0 = table[disp_ram[i]];
    P2 = 0x01<<i;
    delay_ms(1);
    P0 = 0xff;
    P2 = 0;
    }
}
```

```
/*----------------------------------------
  Delay function
  Parameter: unsigned int dt
  Delay time = dt(ms)
---------------------------------------- */
void delay_ms(unsigned int dt)
{
register unsigned char bt,ct;
for (;dt;dt --)
    for (ct = 2;ct;ct --)
        for (bt = 250; -- bt;);
}
```

第 7 章

实验七：LCD 液晶显示器实验

7.1 实验内容与要求

1. 实验目的

学习 AMPIRE128X64 LCD 液晶显示器的字符显示方法。

2. Proteus 仿真实验硬件电路

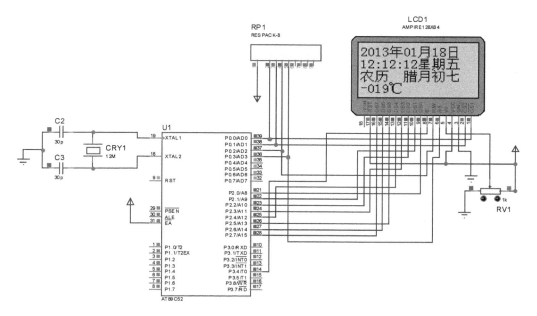

图 7.1 AMPIRE128X64 LCD 液晶显示器实验硬件仿真电路图

3. 实验任务

(1)显示四行汉字与数字。第一行为中文姓名,学号;第二三行为自拟中文文字;第四行为数字或英文。

(2)让数据有动态的变化,或让某个文字闪烁显示。

4. 实验预习要求

(1)学习掌握 AMPIRE128X64 LCD 液晶显示器的线路连接原理,阅读并理解读写函数。

(2)学习掌握汉字的 16×16 点阵及数字、英文字母的 16×8 字模数据产生原理,会用字模软件获取汉字以及数字、英文字母的字模数据。

(3)根据图 7.1 硬件电路原理图,独立画出 Proteus 仿真接线图。

(4)根据实验任务设计编写相应的调试程序。

(5)完成预习报告。

5. 实验设备

计算机(安装单片机汇编、C 编译软件及 Proteus 软件)。

6. 实验报告要求

整理好实验任务 1~2 中经 Proteus 仿真运行正确的程序。

7.2 实验七参考 C 程序

```
/***********************************************/
/*          12864 液晶显示演示 C 程序          */
/*              2013 年 1 月 15 日              */
/***********************************************/
//*************** 预处理 *********************
# include "REG52.H"
# include "lcd12864.h"        //LCD 读写函数
# include "zimo.h"            //点阵字模表
# include "rili.h"            //星期、农历计算函数
# include <stdio.h>           //基本输入输出函数库头文件
# include <math.h>            //数学函数库头文件
# include <stdlib.h>          //标准库头文件
# include <intrins.h>         //左移右移等 C 函数库
```

```
//
#define uchar unsigned char
#define uint unsigned int
//
//******************* 定义 **************************
int temp_val;                    //用以存储数据
uchar CurrentTime_Year = 13,CurrentTime_Month = 1,CurrentTime_Day = 18;//2013 年 1 月 18 日
uchar CurrentTime_Hour = 12,CurrentTime_Minute = 12,CurrentTime_Second = 12;
//12 时 12 分 12 秒
//
//***************** 函数申明 ******************
void Delay1ms(unsigned int); //1ms 延时函数
void display_temp();          //第四行温度显示函数
void show();                  //一到三行显示函数
//
/****************************************/
/*              主函数                  */
/****************************************/
void main()
{
    temp_val = -55;
    iniLCD();                         //初始化 LCD;
while(1)
    {
      show();                         //显示年月日,时分秒,星期,阴历
      display_temp();                 //显示温度
      Delay1ms(500);
      temp_val ++ ;if(temp_val>125)temp_val = -55;
    }
}
//
/****************************************/
/*            延时 1ms 函数              */
/****************************************/
void Delay1ms(unsigned int count)
{
    unsigned int i,j;
    for(i = 0;i<count;i ++ )
    for(j = 0;j<120;j ++ );
```

```
}
//
/***********************************************/
/*     年月日,时分秒,星期,阴历显示函数          */
/***********************************************/
void show()
{      //第一行
    ChangeToLCD(0xb8,0x40,20);                    //表示为 20XX 年,写 20
    ChangeToLCD(0xb8,0x50,CurrentTime_Year);      //显示年数
    display_HZ(0xb8,0x60,nianli[0]);              //显示"年"字
    ChangeToLCD(0xb8,0x70,CurrentTime_Month);     //显示月数
    display_HZ(0xb8,0x80,nianli[1]);              //显示"月"字
    ChangeToLCD(0xb8,0x90,CurrentTime_Day);       //显示日数
    display_HZ(0xb8,0xa0,nianli[2]);              //显示"日"字
    //第二行
    ChangeToLCD(0xba,0x40,CurrentTime_Hour);      //显示时数
    display_FH(0xba,0x50,FH[0]);                  //显示:
    ChangeToLCD(0xba,0x58,CurrentTime_Minute);    //显示分数
    display_FH(0xba,0x68,FH[0]);                  //显示:
    ChangeToLCD(0xba,0x70,CurrentTime_Second);    //显示秒数
    display_HZ(0xba,0x80,WEEK[7]);                //显示"星"字
    display_HZ(0xba,0x90,WEEK[8]);                //显示"期"字
    GetWeek(CurrentTime_Year,CurrentTime_Month,CurrentTime_Day);     //显示星期几
    //第三行
    display_HZ(0xbc,0x40,yinli[0]);               //显示"农"字
    display_HZ(0xbc,0x50,yinli[1]);               //显示"历"字
    Conversion(CurrentTime_Year,CurrentTime_Month,CurrentTime_Day);     //显示阴历
}
//
/***********************************************/
/*       温度显示函数（在第四行）               */
/***********************************************/
void display_temp()
{
    if(temp_val<0)                         //处理温度正负
    {
        display_FH(0xbe,0x40,FH[6]);       //显示温度负
        showtemp(0xbe,0x48,-temp_val);     //显示温度
    }
```

```
        else
    {
            display_FH(0xbe,0x40,FH[5]);        //显示温度正
                showtemp(0xbe,0x48,temp_val);   //显示温度
        }
        display_HZ(0xbe,0x60,sheshidu[0]);      //摄氏度标志"℃"
}
//*************************** C 程序结束 ***************************//
```

以下为程序中用到的头文件，分别是液晶驱动头文件（LCD12864.h）、日历头文件（rili.h）、字模头文件（zimo.h）。

下面为液晶驱动头文件（LCD12864.h）。

```
//****************************************************//
//                                                  //
//              LCD12864 头文件                     //
//              2012 年 1 月 15 日                   //
//****************************************************//
//LCD12864 驱动程序//
//
#ifndef __ LCD12864_H __
#define __ LCD12864_H __
//***************** 预处理 *****************//
#include <intrins.h>
#include"zimo.h"
#define uchar unsigned char
//***************** 端口定义 *****************//
sbit E = P3^4;
sbit RW = P0^2;         //读写控制
sbit RS = P0^3;         //数据指令选择
sbit L = P0^1;          //左屏
sbit R = P0^0;          //右屏
sbit Busy = P2^7;
uchar i,j;
//***************** 函数声明 *****************//
void iniLCD(void);
void chkbusy(void);
void wcode(uchar);
void wdata(uchar);
```

```
void display_HZ(uchar,uchar,uchar * );
void SetOnOff(uchar onoff);
//********************* 延时函数 *************************//
void delay(uchar a)
  {
  uchar i;
  while(a--)
  for(i=100;i>0;i--);
  }
//********************* LCD 写命令 *************************//
void SendCommand(uchar command)
{
    chkbusy();
    E=1;
    RW=0;
    RS=0;
    P2=command;
    E=0;
}
//********************* LCD 显示控制 *************************//
void SetOnOff(uchar onoff)                    //1-开显示 0-关
{
    if(onoff==1)
    {chkbusy();E=1;RW=0;RS=0;P2=0x3f;E=0;}
    else
    P2=0x3e;
  }
//********************* LCD 行设定 *************************//
void SetLine(uchar line)      //line->0：7
{
line=line & 0x07;
line=line | 0xb8;             //1011 1xxx
SendCommand(line);
}
//********************* LCD 色设定 *************************//
void SetColum(uchar colum)   //colum->0：63
{
colum=colum & 0x3f;
colum=colum | 0x40;          //01xx xxxx
```

```c
    SendCommand(colum);
}
//*********************** LCD 屏控制 ***********************//
void SelectScreen(uchar screen) //0-左屏,1-右屏,2-全
{
    switch(screen)
        {
                case 0 :
                L = 0;
                delay(2);
                R = 1;
                delay(2);
                break;
                case 1 :
                L = 1;
                delay(2);
                R = 0;
                delay(2);
                break;
                case 2 :
                L = 0;
                delay(2);
                R = 0;
                delay(2);
                break;
        }
    }
//********************** LCD 清屏 ***********************//
void ClearScreen(uchar screen)
    {
uchar i,j;
SelectScreen(screen);
for(i = 0;i < 8;i++)
{
SetLine(i);
SetColum(0);
for(j = 0;j < 64; j++)
wdata(0x00);
    }
```

```
    }
//********************* LCD 初始化 ***********************//
void iniLCD(void)
{
    wcode(0x3f);        //开显示
    wcode(0xc0);        //显示起始行为第一行
    wcode(0xb8);        //页面地址
    wcode(0x40);        //列地址设为 0
    L = 1;R = 1;
    SetOnOff(1);        //开显示
    ClearScreen(2);     //清屏,点亮背光
}
//********************* LCD 忙检测 ***********************//
void chkbusy(void)
{
    E = 1;              //LCD 使能
    RS = 0;             //读写指令
    RW = 1;             //读状态
    P2 = 0xff;          //P2 输入状态
    while(! Busy);      //忙等待
}
//********************* LCD 写指令 ***********************//
void wcode(uchar cd)
{
    chkbusy();          //等待空闲
    E = 1;              //设置 LCD 写指令状态
    RW = 0;
    RS = 0;
    P2 = cd;            //写指令
    E = 1;
    E = 0;              //产生下降沿
}
//********************* LCD 写数据 ***********************//
void wdata(uchar dat)
{
    chkbusy();          //等待空闲
    E = 1;              //设置 LCD 写数据状态
    RW = 0;
    RS = 1;
```

```
    P2 = dat;                //写数据
    E = 1;
    E = 0;
}
//******************** LCD 显示 16 * 16 汉字程序 ********************//
//
void display_HZ(uchar page,uchar col,uchar  * temp)
{
    L = 1;R = 0;              //从左半屏开始,若列数超过 128,改右
    if(col> = 0x80)
        {
            R = 1;L = 0;
            col - = 0x40;
        }
        wcode(page);              //按要求写入页地址
        wcode(col);              //按要求从相应列开始写数据
        for(j = 0;j<16;j + + )   //写入一个汉字的上半部分,共 16 字节
        {
            wdata(temp[j]);
        }
         wcode(page + 1);//从下一页开始显示汉字的下半部分,要求从相应的列开始写数据
        wcode(col);
        for(j = 16;j<32;j + + )
        {
            wdata(temp[j]);
        }
}
//**************** LCD 显示 16 * 8 英文与数字程序 ****************//
//
void display_FH(uchar page,uchar col,uchar  * temp)       //显示 16 * 8 字符子程序
{
    L = 1;R = 0;                              //从左半屏开始,若列数超过 128,改右
    if(col> = 0x80)
        {
            R = 1;L = 0;
            col - = 0x40;
        }
        wcode(page);                          //按要求写入页地址
        wcode(col);                          //按要求从相应列开始写数据
```

```
    for(j = 0;j<8;j++)
    {
        wdata(temp[j]);
    }
    wcode(page + 1);
    wcode(col);
    for(j = 8;j<16;j++)
    {
        wdata(temp[j]);
    }
}
//******************** 数据(0-255)显示程序 ********************//
void ChangeToLCD(uchar line,uchar column,uchar dat)
{
    int D_ge,D_shi;

    D_ge = dat % 10;                        //取个位
    D_shi = dat % 100/10;                   //取十位

    switch(D_ge)
    {
        case 0:{display_FH(line,column + 8,SZ[0]);break;}
        case 1:{display_FH(line,column + 8,SZ[1]);break;}
        case 2:{display_FH(line,column + 8,SZ[2]);break;}
        case 3:{display_FH(line,column + 8,SZ[3]);break;}
        case 4:{display_FH(line,column + 8,SZ[4]);break;}
        case 5:{display_FH(line,column + 8,SZ[5]);break;}
        case 6:{display_FH(line,column + 8,SZ[6]);break;}
        case 7:{display_FH(line,column + 8,SZ[7]);break;}
        case 8:{display_FH(line,column + 8,SZ[8]);break;}
        case 9:{display_FH(line,column + 8,SZ[9]);break;}
    }
    switch(D_shi)
    {
        case 0:{display_FH(line,column,SZ[0]);break;}
        case 1:{display_FH(line,column,SZ[1]);break;}
        case 2:{display_FH(line,column,SZ[2]);break;}
        case 3:{display_FH(line,column,SZ[3]);break;}
        case 4:{display_FH(line,column,SZ[4]);break;}
```

```
        case 5:{display_FH(line,column,SZ[5]);break;}
        case 6:{display_FH(line,column,SZ[6]);break;}
        case 7:{display_FH(line,column,SZ[7]);break;}
        case 8:{display_FH(line,column,SZ[8]);break;}
        case 9:{display_FH(line,column,SZ[9]);break;}
    }
}
//****************** LCD 显示温度(三位)程序 ********************//
void showtemp(uchar line,uchar column,uchar dat)
{
int D_ge,D_shi,D_bai;
    D_ge = dat % 10;                        //取个位
    D_shi = dat % 100/10;                   //取十位
    D_bai = dat/100;
    switch(D_ge)
    {
        case 0:{display_FH(line,column + 16,SZ[0]);break;}
        case 1:{display_FH(line,column + 16,SZ[1]);break;}
        case 2:{display_FH(line,column + 16,SZ[2]);break;}
        case 3:{display_FH(line,column + 16,SZ[3]);break;}
        case 4:{display_FH(line,column + 16,SZ[4]);break;}
        case 5:{display_FH(line,column + 16,SZ[5]);break;}
        case 6:{display_FH(line,column + 16,SZ[6]);break;}
        case 7:{display_FH(line,column + 16,SZ[7]);break;}
        case 8:{display_FH(line,column + 16,SZ[8]);break;}
        case 9:{display_FH(line,column + 16,SZ[9]);break;}
    }

    switch(D_shi)
    {
        case 0:{display_FH(line,column + 8,SZ[0]);break;}
        case 1:{display_FH(line,column + 8,SZ[1]);break;}
        case 2:{display_FH(line,column + 8,SZ[2]);break;}
        case 3:{display_FH(line,column + 8,SZ[3]);break;}
        case 4:{display_FH(line,column + 8,SZ[4]);break;}
        case 5:{display_FH(line,column + 8,SZ[5]);break;}
        case 6:{display_FH(line,column + 8,SZ[6]);break;}
        case 7:{display_FH(line,column + 8,SZ[7]);break;}
        case 8:{display_FH(line,column + 8,SZ[8]);break;}
        case 9:{display_FH(line,column + 8,SZ[9]);break;}
```

```
    }
    switch(D_bai)
    {
        case 0:{display_FH(line,column,SZ[0]);break;}
        case 1:{display_FH(line,column,SZ[1]);break;}
        case 2:{display_FH(line,column,SZ[2]);break;}
        case 3:{display_FH(line,column,SZ[3]);break;}
        case 4:{display_FH(line,column,SZ[4]);break;}
        case 5:{display_FH(line,column,SZ[5]);break;}
        case 6:{display_FH(line,column,SZ[6]);break;}
        case 7:{display_FH(line,column,SZ[7]);break;}
        case 8:{display_FH(line,column,SZ[8]);break;}
        case 9:{display_FH(line,column,SZ[9]);break;}
    }
}
//*************** LCD 显示星期(一到日)程序 ********************//
void ChangeToLCD4(uchar line,uchar column,uchar dat)
{
    switch(dat)
    {
        case 1:{display_HZ(line,column,WEEK[6]);break;}
        case 2:{display_HZ(line,column,WEEK[0]);break;}
        case 3:{display_HZ(line,column,WEEK[1]);break;}
        case 4:{display_HZ(line,column,WEEK[2]);break;}
        case 5:{display_HZ(line,column,WEEK[3]);break;}
        case 6:{display_HZ(line,column,WEEK[4]);break;}
        case 7:{display_HZ(line,column,WEEK[5]);break;}
    }
}
//************** LCD 显示阴历程序 ********************//
void ChangeToLCD5(uchar line,uchar column,uchar dat)
{
    switch(dat)
    {
        case 0:{display_HZ(line,column,yinli[2]);break;}
        case 1:{display_HZ(line,column,yinli[3]);break;}
        case 2:{display_HZ(line,column,yinli[4]);break;}
        case 3:{display_HZ(line,column,yinli[5]);break;}
        case 4:{display_HZ(line,column,yinli[6]);break;}
```

```
        case 5:{display_HZ(line,column,yinli[7]);break;}
        case 6:{display_HZ(line,column,yinli[8]);break;}
        case 7:{display_HZ(line,column,yinli[9]);break;}
        case 8:{display_HZ(line,column,yinli[10]);break;}
        case 9:{display_HZ(line,column,yinli[11]);break;}
        case 10:{display_HZ(line,column,yinli[12]);break;}
    }
}
#endif
//
/*********************** LCD12864 头文件结束 ***************/
```

下面为日历头文件（rili.h）。

```
//**********************************************//
//                                              //
//                 日历头文件                    //
//               2012 年 1 月 15 日              //
//**********************************************//
//以下为日历计算程序
//
#ifndef __RILI_H__
#define __RILI_H__
//******************** 预处理 ********************//
#define uchar unsigned char
#define uint unsigned int
//
data uchar year_moon,month_moon,day_moon;
bit c_moon;
bit c = 0;
uchar table_week[12] = {0,3,3,6,1,4,6,2,5,0,3,5};
//
//******************** 计算星期程序 ********************//
//
void GetWeek(unsigned int year,unsigned char month,unsigned char day)
{
    unsigned char xingqi;
    unsigned int temp2;
    unsigned char yearH,yearL;
```

```
        yearH = year/100;        yearL = year % 100;


        // 如果为 21 世纪,年份数加 100
        if (yearH>19)                yearL += 100;
        // 所过闰年数只算 1900 年之后的
        temp2 = yearL + yearL/4;
        temp2 = temp2 % 7;
        temp2 = temp2 + day + table_week[month-1];
        if (yearL % 4 == 0&&month<3)    temp2--;  //    根据年份计算星期
        xingqi = (temp2 % 7 + 1);
        ChangeToLCD4(0xba,0xa0,xingqi);
}


//
//********************* 计算农历月的大月或小月 **************************//
//如果该月为大返回 1,为小返回 0
//
bit get_moon_day(uchar month_p,uint table_addr)
{
uchar temp;
switch (month_p)
{
case 1:{temp = year_code[table_addr]&0x08;
if (temp == 0)return(0);else return(1);}
case 2:{temp = year_code[table_addr]&0x04;
if (temp == 0)return(0);else return(1);}
case 3:{temp = year_code[table_addr]&0x02;
if (temp == 0)return(0);else return(1);}
case 4:{temp = year_code[table_addr]&0x01;
if (temp == 0)return(0);else return(1);}
case 5:{temp = year_code[table_addr + 1]&0x80;
if (temp == 0) return(0);else return(1);}
case 6:{temp = year_code[table_addr + 1]&0x40;
if (temp == 0)return(0);else return(1);}
case 7:{temp = year_code[table_addr + 1]&0x20;
if (temp == 0)return(0);else return(1);}
case 8:{temp = year_code[table_addr + 1]&0x10;
if (temp == 0)return(0);else return(1);}
case 9:{temp = year_code[table_addr + 1]&0x08;
```

```
      if(temp==0)return(0);else return(1);}
      case 10:{temp=year_code[table_addr+1]&0x04;
      if(temp==0)return(0);else return(1);}
      case 11:{temp=year_code[table_addr+1]&0x02;
      if(temp==0)return(0);else return(1);}
      case 12:{temp=year_code[table_addr+1]&0x01;
      if(temp==0)return(0);else return(1);}
      case 13:{temp=year_code[table_addr+2]&0x80;
      if(temp==0)return(0);else return(1);}
      }
  }
//
//********************* 计算农历年月日程序 *********************//
/*  函数功能:输入 BCD 阳历数据,输出 BCD 阴历数据(只算 2000 - 2099 年)
如:计算 2012 年 1 月 16 日为　　Conversion(0,0x12,0x01,0x16);
调用函数后,原有数据不变,读 c_moon,year_moon,month_moon,day_moon 得出阴历 BCD 数据  */
//
void Conversion(uchar year,uchar month,uchar day)
{
uchar temp1,temp2,temp3,month_p;
uint temp4,table_addr;
bit flag2,flag_y;
table_addr=(year+0x64-1)*0x3;
//
//定位数据表地址完成
//取当年春节所在的公历月份
temp1=year_code[table_addr+2]&0x60;
temp1=_cror_(temp1,5);
//取当年春节所在的公历月份完成
//取当年春节所在的公历日
temp2=year_code[table_addr+2]&0x1f;
//取当年春节所在的公历日完成
//计算当年春年离当年元旦的天数,春节只会在公历 1 月或 2 月
if(temp1==0x1)
{
temp3=temp2-1;
}
else
{
```

```
temp3 = temp2 + 0x1f - 1;
}
//计算当年春年离当年元旦的天数完成
//计算公历日离当年元旦的天数,为了减少运算,用了两个表
//day_code1[9],day_code2[3]
//如果公历月在九月或前,天数会少于 0xff,用表 day_code1[9]
//在九月后,天数大于 0xff,用表 day_code2[3]
//如输入公历日为 8 月 10 日,则公历日离元旦天数为 day_code1[8 - 1] + 10 - 1
//如输入公历日为 11 月 10 日,则公历日离元旦天数为 day_code2[11 - 10] + 10 - 1
if (month<10)
{
temp4 = day_code1[month - 1] + day - 1;
}
else
{
temp4 = day_code2[month - 10] + day - 1;
}
if ((month>0x2)&&(year % 0x4 ==  0))
{ //如果公历月大于 2 月并且该年的 2 月为闰月,天数加 1
temp4 += 1;
}
//计算公历日离当年元旦的天数完成
//判断公历日在春节前还是春节后
if (temp4> = temp3)
{ //公历日在春节后或就是春节当日使用下面代码进行运算
temp4 -= temp3;
month = 0x1;
month_p = 0x1; //month_p 为月份指向,公历日在春节前或就是春节当日 month_p 指向首月
flag2 = get_moon_day(month_p,table_addr);
//检查该农历月为大小还是小月,大月返回 1,小月返回 0
flag_y = 0;
if(flag2 ==  0)temp1 = 0x1d; //小月 29 天
else temp1 = 0x1e; //大小 30 天
temp2 = year_code[table_addr]&0xf0;
temp2 = _cror_(temp2,4); //从数据表中取该年的闰月月份,如为 0 则该年无闰月
while(temp4> = temp1)
{
temp4 -= temp1;
month_p += 1;
```

```
if(month == temp2)
{
flag_y = ~flag_y;
if(flag_y == 0)
month += 1;
}
else month += 1;
flag2 = get_moon_day(month_p,table_addr);
if(flag2 == 0)temp1 = 0x1d;
else temp1 = 0x1e;
}
day = temp4 + 1;
}
else
{ //公历日在春节前使用下面代码进行运算
temp3 -= temp4;
if (year == 0x0)
{
year = 0x63;c = 1;
}
else year -= 1;
table_addr -= 0x3;
month = 0xc;
temp2 = year_code[table_addr]&0xf0;
temp2 = _cror_(temp2,4);
if (temp2 == 0)
month_p = 0xc;
else
month_p = 0xd; //
/* month_p 为月份指向,如果当年有闰月,一年有十三个月,月指向13,无闰月指向12 */
flag_y = 0;
flag2 = get_moon_day(month_p,table_addr);
if(flag2 == 0)temp1 = 0x1d;
else temp1 = 0x1e;
    while(temp3>temp1)
    {
        temp3 -= temp1;
        month_p -= 1;
        if(flag_y == 0)month -= 1;
```

```
            if(month == temp2)flag_y = ~flag_y;
            flag2 = get_moon_day(month_p,table_addr);
            if(flag2 == 0)temp1 = 0x1d;
            else temp1 = 0x1e;
        }
    day = temp1 - temp3 + 1;
}


/* 显示阴历的月份 */
    switch(month)
    {
        case 1:{display_HZ(0xbc,0x60,hei[0]);display_HZ(0xbc,0x70,yinli[13]);break;}
        case 2:
        case 3:
        case 4:
        case 5:
        case 6:
        case 7:
        case 8:
        case 9:{display_HZ(0xbc,0x60,hei[0]);ChangeToLCD5(0xbc,0x70,month); break;}
//      case 10:{display_HZ(0xbc,0x60,hei[0]);display_HZ(0xbc,0x80,yinli[12]);break; }
        case 10:{display_HZ(0xbc,0x60,hei[0]);display_HZ(0xbc,0x70,yinli[12]);break; };
        case 11:{display_HZ(0xbc,0x60,yinli[12]);display_HZ(0xbc,0x70,yinli[3]);break;}
        case 12:{display_HZ(0xbc,0x60,hei[0]);display_HZ(0xbc,0x70,yinli[14]);break;}
        break;
    }
    display_HZ(0xbc,0x80,nianli[1]);
/* 显示阴历的日子 */
    temp1 = day/10;
    temp2 = day%10;
            switch(day)
    {
        case 1:
        case 2:
        case 3:
        case 4:
        case 5:
        case 6:
        case 7:
```

```
        case 8：
        case 9：
        case 10：{display_HZ(0xbc,0x90,yinli[2]);ChangeToLCD5(0xbc,0xa0,day)；break;}
        case 11：
        case 12：
        case 13：
        case 14：
        case 15：
        case 16：
        case 17：
        case 18：
        case 19：
            {display_HZ(0xbc,0x90,yinli[12]);ChangeToLCD5(0xbc,0xa0,temp2);
            break;}
        case 20：{display_HZ(0xbc,0x90,yinli[4]);display_HZ(0xbc,0xa0,yinli[12]);break;}
        case 21：
        case 22：
        case 23：
        case 24：
        case 25：
        case 26：
        case 27：
        case 28：
        case 29：{display_HZ(0xbc,0x90,yinli[16]);ChangeToLCD5(0xbc,0xa0,temp2);break;}
        case 30：{display_HZ(0xbc,0x90,yinli[5]);display_HZ(0xbc,0xa0,yinli[12]);break;}
    }
}
＃endif
//
/*********************** 日历头文件结束 ********************************/
```

下面为字模头文件(zimo.h)。

```
        /**********************************************/
        /*        12864 液晶显示用中西文字模表        */
        /*              2013 年 1 月 15 日            */
        /**********************************************/
＃ifndef __ ZIMO_H__
＃define __ ZIMO_H__
```

```
//
#define uchar unsigned char
#define uint unsigned int
/*******************************************************************
公历年对应的农历数据,每年三字节,
格式第一字节 BIT7－4 位表示闰月月份,值为 0 为无闰月,BIT3－0 对应农历第 1－4 月的大小
第二字节 BIT7－0 对应农历第 5－12 月大小,第三字节 BIT7 表示农历第 13 个月大小
月份对应的位为 1 表示本农历月大(30 天),为 0 表示小(29 天)
第三字节 BIT6－5 表示春节的公历月份,BIT4－0 表示春节的公历日期
/*******************************************************************
code uchar year_code[597] = {
0x04,0xAe,0x53, //1901 0
0x0A,0x57,0x48, //1902 3
0x55,0x26,0xBd, //1903 6
0x0d,0x26,0x50, //1904 9
0x0d,0x95,0x44, //1905 12
0x46,0xAA,0xB9, //1906 15
0x05,0x6A,0x4d, //1907 18
0x09,0xAd,0x42, //1908 21
0x24,0xAe,0xB6, //1909
0x04,0xAe,0x4A, //1910
0x6A,0x4d,0xBe, //1911
0x0A,0x4d,0x52, //1912
0x0d,0x25,0x46, //1913
0x5d,0x52,0xBA, //1914
0x0B,0x54,0x4e, //1915
0x0d,0x6A,0x43, //1916
0x29,0x6d,0x37, //1917
0x09,0x5B,0x4B, //1918
0x74,0x9B,0xC1, //1919
0x04,0x97,0x54, //1920
0x0A,0x4B,0x48, //1921
0x5B,0x25,0xBC, //1922
0x06,0xA5,0x50, //1923
0x06,0xd4,0x45, //1924
0x4A,0xdA,0xB8, //1925
0x02,0xB6,0x4d, //1926
0x09,0x57,0x42, //1927
0x24,0x97,0xB7, //1928
```

```
0x04,0x97,0x4A, //1929
0x66,0x4B,0x3e, //1930
0x0d,0x4A,0x51, //1931
0x0e,0xA5,0x46, //1932
0x56,0xd4,0xBA, //1933
0x05,0xAd,0x4e, //1934
0x02,0xB6,0x44, //1935
0x39,0x37,0x38, //1936
0x09,0x2e,0x4B, //1937
0x7C,0x96,0xBf, //1938
0x0C,0x95,0x53, //1939
0x0d,0x4A,0x48, //1940
0x6d,0xA5,0x3B, //1941
0x0B,0x55,0x4f, //1942
0x05,0x6A,0x45, //1943
0x4A,0xAd,0xB9, //1944
0x02,0x5d,0x4d, //1945
0x09,0x2d,0x42, //1946
0x2C,0x95,0xB6, //1947
0x0A,0x95,0x4A, //1948
0x7B,0x4A,0xBd, //1949
0x06,0xCA,0x51, //1950
0x0B,0x55,0x46, //1951
0x55,0x5A,0xBB, //1952
0x04,0xdA,0x4e, //1953
0x0A,0x5B,0x43, //1954
0x35,0x2B,0xB8, //1955
0x05,0x2B,0x4C, //1956
0x8A,0x95,0x3f, //1957
0x0e,0x95,0x52, //1958
0x06,0xAA,0x48, //1959
0x7A,0xd5,0x3C, //1960
0x0A,0xB5,0x4f, //1961
0x04,0xB6,0x45, //1962
0x4A,0x57,0x39, //1963
0x0A,0x57,0x4d, //1964
0x05,0x26,0x42, //1965
0x3e,0x93,0x35, //1966
0x0d,0x95,0x49, //1967
```

```
0x75,0xAA,0xBe, //1968
0x05,0x6A,0x51, //1969
0x09,0x6d,0x46, //1970
0x54,0xAe,0xBB, //1971
0x04,0xAd,0x4f, //1972
0x0A,0x4d,0x43, //1973
0x4d,0x26,0xB7, //1974
0x0d,0x25,0x4B, //1975
0x8d,0x52,0xBf, //1976
0x0B,0x54,0x52, //1977
0x0B,0x6A,0x47, //1978
0x69,0x6d,0x3C, //1979
0x09,0x5B,0x50, //1980
0x04,0x9B,0x45, //1981
0x4A,0x4B,0xB9, //1982
0x0A,0x4B,0x4d, //1983
0xAB,0x25,0xC2, //1984
0x06,0xA5,0x54, //1985
0x06,0xd4,0x49, //1986
0x6A,0xdA,0x3d, //1987
0x0A,0xB6,0x51, //1988
0x09,0x37,0x46, //1989
0x54,0x97,0xBB, //1990
0x04,0x97,0x4f, //1991
0x06,0x4B,0x44, //1992
0x36,0xA5,0x37, //1993
0x0e,0xA5,0x4A, //1994
0x86,0xB2,0xBf, //1995
0x05,0xAC,0x53, //1996
0x0A,0xB6,0x47, //1997
0x59,0x36,0xBC, //1998
0x09,0x2e,0x50, //1999 294
0x0C,0x96,0x45, //2000 297
0x4d,0x4A,0xB8, //2001
0x0d,0x4A,0x4C, //2002
0x0d,0xA5,0x41, //2003
0x25,0xAA,0xB6, //2004
0x05,0x6A,0x49, //2005
0x7A,0xAd,0xBd, //2006
```

```
0x02,0x5d,0x52, //2007
0x09,0x2d,0x47, //2008
0x5C,0x95,0xBA, //2009
0x0A,0x95,0x4e, //2010
0x0B,0x4A,0x43, //2011
0x4B,0x55,0x37, //2012
0x0A,0xd5,0x4A, //2013
0x95,0x5A,0xBf, //2014
0x04,0xBA,0x53, //2015
0x0A,0x5B,0x48, //2016
0x65,0x2B,0xBC, //2017
0x05,0x2B,0x50, //2018
0x0A,0x93,0x45, //2019
0x47,0x4A,0xB9, //2020
0x06,0xAA,0x4C, //2021
0x0A,0xd5,0x41, //2022
0x24,0xdA,0xB6, //2023
0x04,0xB6,0x4A, //2024
0x69,0x57,0x3d, //2025
0x0A,0x4e,0x51, //2026
0x0d,0x26,0x46, //2027
0x5e,0x93,0x3A, //2028
0x0d,0x53,0x4d, //2029
0x05,0xAA,0x43, //2030
0x36,0xB5,0x37, //2031
0x09,0x6d,0x4B, //2032
0xB4,0xAe,0xBf, //2033
0x04,0xAd,0x53, //2034
0x0A,0x4d,0x48, //2035
0x6d,0x25,0xBC, //2036
0x0d,0x25,0x4f, //2037
0x0d,0x52,0x44, //2038
0x5d,0xAA,0x38, //2039
0x0B,0x5A,0x4C, //2040
0x05,0x6d,0x41, //2041
0x24,0xAd,0xB6, //2042
0x04,0x9B,0x4A, //2043
0x7A,0x4B,0xBe, //2044
0x0A,0x4B,0x51, //2045
```

```
0x0A,0xA5,0x46, //2046
0x5B,0x52,0xBA, //2047
0x06,0xd2,0x4e, //2048
0x0A,0xdA,0x42, //2049
0x35,0x5B,0x37, //2050
0x09,0x37,0x4B, //2051
0x84,0x97,0xC1, //2052
0x04,0x97,0x53, //2053
0x06,0x4B,0x48, //2054
0x66,0xA5,0x3C, //2055
0x0e,0xA5,0x4f, //2056
0x06,0xB2,0x44, //2057
0x4A,0xB6,0x38, //2058
0x0A,0xAe,0x4C, //2059
0x09,0x2e,0x42, //2060
0x3C,0x97,0x35, //2061
0x0C,0x96,0x49, //2062
0x7d,0x4A,0xBd, //2063
0x0d,0x4A,0x51, //2064
0x0d,0xA5,0x45, //2065
0x55,0xAA,0xBA, //2066
0x05,0x6A,0x4e, //2067
0x0A,0x6d,0x43, //2068
0x45,0x2e,0xB7, //2069
0x05,0x2d,0x4B, //2070
0x8A,0x95,0xBf, //2071
0x0A,0x95,0x53, //2072
0x0B,0x4A,0x47, //2073
0x6B,0x55,0x3B, //2074
0x0A,0xd5,0x4f, //2075
0x05,0x5A,0x45, //2076
0x4A,0x5d,0x38, //2077
0x0A,0x5B,0x4C, //2078
0x05,0x2B,0x42, //2079
0x3A,0x93,0xB6, //2080
0x06,0x93,0x49, //2081
0x77,0x29,0xBd, //2082
0x06,0xAA,0x51, //2083
0x0A,0xd5,0x46, //2084
```

```
0x54,0xdA,0xBA, //2085
0x04,0xB6,0x4e, //2086
0x0A,0x57,0x43, //2087
0x45,0x27,0x38, //2088
0x0d,0x26,0x4A, //2089
0x8e,0x93,0x3e, //2090
0x0d,0x52,0x52, //2091
0x0d,0xAA,0x47, //2092
0x66,0xB5,0x3B, //2093
0x05,0x6d,0x4f, //2094
0x04,0xAe,0x45, //2095
0x4A,0x4e,0xB9, //2096
0x0A,0x4d,0x4C, //2097
0x0d,0x15,0x41, //2098
0x2d,0x92,0xB5, //2099
};
```

//**************** 月份数据表 *********************************
```
code uchar day_code1[9] = {0x0,0x1f,0x3b,0x5a,0x78,0x97,0xb5,0xd4,0xf3};
code uint day_code2[3] = {0x111,0x130,0x14e};
```
//**************** 符号数据表 *********************************
```
/* -- 此字体下对应的点阵为:宽×高=8×16  -- */
const uchar code FH[][16] = {                    //符号

    0x00,0x00,0x00,0xC0,0xC0,0x00,0x00,0x00, //：
    0x00,0x00,0x00,0x30,0x30,0x00,0x00,0x00,

    0x00,0x00,0x00,0x00,0x00,0x00,0x00,0x00, //.
    0x00,0x30,0x30,0x00,0x00,0x00,0x00,0x00,

    0x08,0xF8,0x00,0x00,0x80,0x80,0x80,0x00, //k
    0x20,0x3F,0x24,0x02,0x2D,0x30,0x20,0x00,

    0x80,0x80,0x80,0x80,0x80,0x80,0x80,0x00, //m
    0x20,0x3F,0x20,0x00,0x3F,0x20,0x00,0x3F,

    0x00,0x80,0x80,0x80,0x80,0x80,0x80,0x00,
    0x00,0x00,0x00,0x00,0x00,0x00,0x00,0x00, //-

    0x00,0x80,0x80,0xE0,0xE0,0x80,0x80,0x00,
```

```
0x00,0x00,0x00,0x03,0x03,0x00,0x00,0x00, //+

0x00,0x80,0x80,0x80,0x80,0x80,0x80,0x00,
0x00,0x00,0x00,0x00,0x00,0x00,0x00,0x00, //-

                                  };
//**************** 0-9 数字字模表 ****************************************
/*--   此字体下对应的点阵为:宽×高 = 8×16   --*/
const uchar code SZ[][16] = {                         //数字 ;
    0x00,0xE0,0x10,0x08,0x08,0x10,0xE0,0x00,//0
    0x00,0x0F,0x10,0x20,0x20,0x10,0x0F,0x00,

    0x00,0x10,0x10,0xF8,0x00,0x00,0x00,0x00,//1
    0x00,0x20,0x20,0x3F,0x20,0x20,0x00,0x00,

    0x00,0x70,0x08,0x08,0x08,0x88,0x70,0x00,//2
    0x00,0x30,0x28,0x24,0x22,0x21,0x30,0x00,

    0x00,0x30,0x08,0x88,0x88,0x48,0x30,0x00,//3
    0x00,0x18,0x20,0x20,0x20,0x11,0x0E,0x00,

    0x00,0x00,0xC0,0x20,0x10,0xF8,0x00,0x00,//4
    0x00,0x07,0x04,0x24,0x24,0x3F,0x24,0x00,

    0x00,0xF8,0x08,0x88,0x88,0x08,0x08,0x00,//5
    0x00,0x19,0x21,0x20,0x20,0x11,0x0E,0x00,

    0x00,0xE0,0x10,0x88,0x88,0x18,0x00,0x00,//6
    0x00,0x0F,0x11,0x20,0x20,0x11,0x0E,0x00,

    0x00,0x38,0x08,0x08,0xC8,0x38,0x08,0x00,//7
    0x00,0x00,0x00,0x3F,0x00,0x00,0x00,0x00,

    0x00,0x70,0x88,0x08,0x08,0x88,0x70,0x00,//8
    0x00,0x1C,0x22,0x21,0x21,0x22,0x1C,0x00,


    0x00,0xE0,0x10,0x08,0x08,0x10,0xE0,0x00,//9
    0x00,0x00,0x31,0x22,0x22,0x11,0x0F,0x00,
```

```
                                    };
//*******************************************************
const uchar code hei[][32]  =
{

    0x00,0x00,0x00,0x00,0x00,0x00,0x00,0x00,0x00,0x00,0x00,0x00, 0x00,0x00,0x00,0x00,
    0x00,0x00,0x00,0x00,0x00,0x00,0x00,0x00,0x00,0x00,0x00,0x00, 0x00,0x00,0x00,0x00,
};
//*************** 中文字符数据表 *******************************
const uchar code DAXIE[][32] = {

/*--  文字：  零 --*/
/*-- Fixedsys12；  此字体下对应的点阵为:宽×高 = 16×16   --*/
0x10,0x0C,0x05,0x55,0x55,0x55,0x85,0x7F,0x85,0x55,0x55,0x55,0x05,0x14,0x0C,0x00,
0x04,0x04,0x02,0x0A,0x09,0x29,0x2A,0x4C,0x48,0xA9,0x19,0x02,0x02,0x04,0x04,0x00,

/*--  文字：  一 --*/
/*-- Fixedsys12；  此字体下对应的点阵为:宽×高 = 16×16   --*/
0x80,0x80,0x80,0x80,0x80,0x80,0x80,0x80,0x80,0x80,0x80,0x80,0x80,0x80,0x80,0x00,
0x00,0x00,0x00,0x00,0x00,0x00,0x00,0x00,0x00,0x00,0x00,0x00,0x00,0x00,0x00,0x00,

/*--  文字：  二 --*/
/*-- Fixedsys12；  此字体下对应的点阵为:宽×高 = 16×16   --*/
0x00,0x00,0x08,0x08,0x08,0x08,0x08,0x08,0x08,0x08,0x08,0x08,0x08,0x00,0x00,0x00,
0x10,0x10,0x10,0x10,0x10,0x10,0x10,0x10,0x10,0x10,0x10,0x10,0x10,0x10,0x10,0x00,

/*--  文字：  三 --*/
/*-- Fixedsys12；  此字体下对应的点阵为:宽×高 = 16×16   --*/
0x00,0x04,0x84,0x84,0x84,0x84,0x84,0x84,0x84,0x84,0x84,0x84,0x84,0x04,0x00,0x00,
0x20,0x20,0x20,0x20,0x20,0x20,0x20,0x20,0x20,0x20,0x20,0x20,0x20,0x20,0x20,0x00,

/*--  文字：  四 --*/
/*-- Fixedsys12；  此字体下对应的点阵为:宽×高 = 16×16   --*/
0x00,0xFC,0x04,0x04,0x04,0xFC,0x04,0x04,0x04,0xFC,0x04,0x04,0x04,0xFC,0x00,0x00,
0x00,0x7F,0x28,0x24,0x23,0x20,0x20,0x20,0x20,0x21,0x22,0x22,0x22,0x7F,0x00,0x00,

/*--  文字：  五 --*/
/*-- Fixedsys12；  此字体下对应的点阵为:宽×高 = 16×16   --*/
0x00,0x02,0x42,0x42,0x42,0xC2,0x7E,0x42,0x42,0x42,0x42,0xC2,0x02,0x02,0x00,0x00,
```

```
0x40,0x40,0x40,0x40,0x78,0x47,0x40,0x40,0x40,0x40,0x40,0x7F,0x40,0x40,0x40,0x00,


/*-- 文字：  六 --*/
/*-- Fixedsys12；  此字体下对应的点阵为:宽×高 = 16×16   --*/
0x20,0x20,0x20,0x20,0x20,0x20,0x21,0x22,0x2C,0x20,0x20,0x20,0x20,0x20,0x20,0x00,
0x00,0x40,0x20,0x10,0x0C,0x03,0x00,0x00,0x00,0x01,0x02,0x04,0x18,0x60,0x00,0x00,


/*-- 文字：  七 --*/
/*-- Fixedsys12；  此字体下对应的点阵为:宽×高 = 16×16   --*/
0x80,0x80,0x80,0x80,0x80,0x40,0xFF,0x40,0x40,0x40,0x20,0x20,0x20,0x20,0x00,0x00,
0x00,0x00,0x00,0x00,0x00,0x00,0x3F,0x40,0x40,0x40,0x40,0x40,0x40,0x78,0x00,0x00,


/*-- 文字：  八 --*/
/*-- Fixedsys12；  此字体下对应的点阵为:宽×高 = 16×16   --*/
0x00,0x00,0x00,0x00,0x00,0xFC,0x00,0x00,0x00,0x7E,0x80,0x00,0x00,0x00,0x00,0x00,
0x00,0x80,0x60,0x18,0x07,0x00,0x00,0x00,0x00,0x00,0x03,0x0C,0x30,0x40,0x80,0x00,


/*-- 文字：  九 --*/
/*-- Fixedsys12；  此字体下对应的点阵为:宽×高 = 16×16   --*/
0x00,0x10,0x10,0x10,0x10,0xFF,0x10,0x10,0x10,0x10,0xF0,0x00,0x00,0x00,0x00,0x00,
0x80,0x40,0x20,0x18,0x07,0x00,0x00,0x00,0x00,0x00,0x3F,0x40,0x40,0x40,0x78,0x00,
};
//*************** 星期数据表 *****************************************
const uchar code WEEK[][32] =
{


/*-- 文字：  日 --*/
/*-- Fixedsys12；  此字体下对应的点阵为:宽×高 = 16×16   --*/
0x00,0x00,0x00,0xFE,0x82,0x82,0x82,0x82,0x82,0x82,0x82,0xFE,0x00,0x00,0x00,0x00,
0x00,0x00,0x00,0xFF,0x40,0x40,0x40,0x40,0x40,0x40,0x40,0xFF,0x00,0x00,0x00,0x00,


/*-- 文字：  一 --*/
/*-- Fixedsys12；  此字体下对应的点阵为:宽×高 = 16×16   --*/
0x80,0x80,0x80,0x80,0x80,0x80,0x80,0x80,0x80,0x80,0x80,0x80,0x80,0x80,0x80,0x00,
0x00,0x00,0x00,0x00,0x00,0x00,0x00,0x00,0x00,0x00,0x00,0x00,0x00,0x00,0x00,0x00,


/*-- 文字：  二 --*/
/*-- Fixedsys12；  此字体下对应的点阵为:宽×高 = 16×16   --*/
0x00,0x00,0x08,0x08,0x08,0x08,0x08,0x08,0x08,0x08,0x08,0x08,0x08,0x00,0x00,0x00,
```

0x10,0x10,0x10,0x10,0x10,0x10,0x10,0x10,0x10,0x10,0x10,0x10,0x10,0x10,0x10,0x00,

/*-- 文字：　三　--*/
/*-- Fixedsys12；　此字体下对应的点阵为:宽×高＝16×16　　--*/
0x00,0x04,0x84,0x84,0x84,0x84,0x84,0x84,0x84,0x84,0x84,0x84,0x04,0x00,0x00,
0x20,0x20,0x20,0x20,0x20,0x20,0x20,0x20,0x20,0x20,0x20,0x20,0x20,0x20,0x20,0x00,

/*-- 文字：　四　--*/
/*-- Fixedsys12；　此字体下对应的点阵为:宽×高＝16×16　　--*/
0x00,0xFC,0x04,0x04,0x04,0xFC,0x04,0x04,0x04,0xFC,0x04,0x04,0x04,0xFC,0x00,0x00,
0x00,0x7F,0x28,0x24,0x23,0x20,0x20,0x20,0x20,0x21,0x22,0x22,0x22,0x7F,0x00,0x00,

/*-- 文字：　五　--*/
/*-- Fixedsys12；　此字体下对应的点阵为:宽×高＝16×16　　--*/
0x00,0x02,0x42,0x42,0x42,0xC2,0x7E,0x42,0x42,0x42,0x42,0xC2,0x02,0x02,0x00,0x00,
0x40,0x40,0x40,0x40,0x78,0x47,0x40,0x40,0x40,0x40,0x40,0x7F,0x40,0x40,0x40,0x00,

/*-- 文字：　六　--*/
/*-- Fixedsys12；　此字体下对应的点阵为:宽×高＝16×16　　--*/
0x20,0x20,0x20,0x20,0x20,0x20,0x21,0x22,0x2C,0x20,0x20,0x20,0x20,0x20,0x20,0x00,
0x00,0x40,0x20,0x10,0x0C,0x03,0x00,0x00,0x00,0x01,0x02,0x04,0x18,0x60,0x00,0x00,

/*-- 文字：　星　--*/
/*-- Fixedsys12；　此字体下对应的点阵为:宽×高＝16×16　　--*/
0x00,0x00,0x00,0xBE,0x2A,0x2A,0x2A,0xEA,0x2A,0x2A,0x2A,0x3E,0x00,0x00,0x00,0x00,
0x00,0x44,0x42,0x49,0x49,0x49,0x49,0x7F,0x49,0x49,0x49,0x49,0x41,0x40,0x00,0x00,

/*-- 文字：　期　--*/
/*-- Fixedsys12；　此字体下对应的点阵为:宽×高＝16×16　　--*/
0x00,0x04,0xFF,0x24,0x24,0x24,0xFF,0x04,0x00,0xFE,0x22,0x22,0x22,0xFE,0x00,0x00,
0x88,0x48,0x2F,0x09,0x09,0x19,0xAF,0x48,0x30,0x0F,0x02,0x42,0x82,0x7F,0x00,0x00,

};
//＊＊＊＊＊＊＊＊＊＊＊＊＊＊＊ 年月日数据表 ＊＊＊＊＊＊＊＊＊＊＊＊＊＊＊＊＊＊＊＊＊＊＊＊＊＊＊
const uchar code nianli[][32] =
{
/*-- 文字：　年　--*/
/*-- Fixedsys12；　此字体下对应的点阵为:宽×高＝16×16　　--*/
0x00,0x20,0x18,0xC7,0x44,0x44,0x44,0x44,0xFC,0x44,0x44,0x44,0x44,0x04,0x00,0x00,

0x04,0x04,0x04,0x07,0x04,0x04,0x04,0x04,0xFF,0x04,0x04,0x04,0x04,0x04,0x04,0x00,

```
/*-- 文字：  月  --*/
/*-- Fixedsys12；  此字体下对应的点阵为:宽×高 = 16×16    --*/
0x00,0x00,0x00,0xFE,0x22,0x22,0x22,0x22,0x22,0x22,0x22,0x22,0xFE,0x00,0x00,0x00,
0x80,0x40,0x30,0x0F,0x02,0x02,0x02,0x02,0x02,0x02,0x42,0x82,0x7F,0x00,0x00,0x00,

/*-- 文字：  日  --*/
/*-- Fixedsys12；  此字体下对应的点阵为:宽×高 = 16×16    --*/
0x00,0x00,0x00,0xFE,0x82,0x82,0x82,0x82,0x82,0x82,0x82,0xFE,0x00,0x00,0x00,0x00,
0x00,0x00,0x00,0xFF,0x40,0x40,0x40,0x40,0x40,0x40,0x40,0xFF,0x00,0x00,0x00,0x00,
};


//***************** 农历汉字数据表 ******************************************
const uchar code yinli[][32] =
{
/*-- 文字：  农  --*/
/*-- Fixedsys12；  此字体下对应的点阵为:宽×高 = 16×16    --*/
0x20,0x18,0x08,0x08,0x08,0xC8,0x38,0xCF,0x08,0x08,0x08,0x08,0xA8,0x18,0x00,0x00,
0x10,0x08,0x04,0x02,0xFF,0x40,0x20,0x00,0x03,0x04,0x0A,0x11,0x20,0x40,0x40,0x00,

/*-- 文字：  历  --*/
/*-- Fixedsys12；  此字体下对应的点阵为:宽×高 = 16×16    --*/
0x00,0x00,0xFE,0x02,0x42,0x42,0x42,0x42,0xFA,0x42,0x42,0x42,0x42,0xC2,0x02,0x00,
0x80,0x60,0x1F,0x80,0x40,0x20,0x18,0x06,0x01,0x00,0x40,0x80,0x40,0x3F,0x00,0x00,

/*-- 文字：  初  --*/
/*-- Fixedsys12；  此字体下对应的点阵为:宽×高 = 16×16    --*/
0x08,0x08,0x89,0xEA,0x18,0x88,0x00,0x04,0x04,0xFC,0x04,0x04,0x04,0xFC,0x00,0x00,
0x02,0x01,0x00,0xFF,0x01,0x86,0x40,0x20,0x18,0x07,0x40,0x80,0x40,0x3F,0x00,0x00,

/*-- 文字：  一  --*/
/*-- Fixedsys12；  此字体下对应的点阵为:宽×高 = 16×16    --*/
0x80,0x80,0x80,0x80,0x80,0x80,0x80,0x80,0x80,0x80,0x80,0x80,0x80,0x80,0x80,0x00,
0x00,0x00,0x00,0x00,0x00,0x00,0x00,0x00,0x00,0x00,0x00,0x00,0x00,0x00,0x00,0x00,

/*-- 文字：  二  --*/
/*-- Fixedsys12；  此字体下对应的点阵为:宽×高 = 16×16    --*/
0x00,0x00,0x08,0x08,0x08,0x08,0x08,0x08,0x08,0x08,0x08,0x08,0x08,0x00,0x00,0x00,
```

0x10,0x10,0x10,0x10,0x10,0x10,0x10,0x10,0x10,0x10,0x10,0x10,0x10,0x10,0x10,0x00,

/*-- 文字：　三 --*/
/*-- Fixedsys12；　此字体下对应的点阵为:宽×高 = 16×16　 --*/
0x00,0x04,0x84,0x84,0x84,0x84,0x84,0x84,0x84,0x84,0x84,0x84,0x84,0x04,0x00,0x00,
0x20,0x20,0x20,0x20,0x20,0x20,0x20,0x20,0x20,0x20,0x20,0x20,0x20,0x20,0x20,0x00,

/*-- 文字：　四 --*/
/*-- Fixedsys12；　此字体下对应的点阵为:宽×高 = 16×16　 --*/
0x00,0xFC,0x04,0x04,0x04,0xFC,0x04,0x04,0x04,0xFC,0x04,0x04,0x04,0xFC,0x00,0x00,
0x00,0x7F,0x28,0x24,0x23,0x20,0x20,0x20,0x20,0x21,0x22,0x22,0x22,0x7F,0x00,0x00,

/*-- 文字：　五 --*/
/*-- Fixedsys12；　此字体下对应的点阵为:宽×高 = 16×16　 --*/
0x00,0x02,0x42,0x42,0x42,0xC2,0x7E,0x42,0x42,0x42,0x42,0xC2,0x02,0x02,0x00,0x00,
0x40,0x40,0x40,0x40,0x78,0x47,0x40,0x40,0x40,0x40,0x40,0x7F,0x40,0x40,0x40,0x00,

/*-- 文字：　六 --*/
/*-- Fixedsys12；　此字体下对应的点阵为:宽×高 = 16×16　 --*/
0x20,0x20,0x20,0x20,0x20,0x20,0x21,0x22,0x2C,0x20,0x20,0x20,0x20,0x20,0x20,0x00,
0x00,0x40,0x20,0x10,0x0C,0x03,0x00,0x00,0x00,0x01,0x02,0x04,0x18,0x60,0x00,0x00,

/*-- 文字：　七 --*/
/*-- Fixedsys12；　此字体下对应的点阵为:宽×高 = 16×16　 --*/
0x80,0x80,0x80,0x80,0x80,0x40,0xFF,0x40,0x40,0x40,0x20,0x20,0x20,0x20,0x00,0x00,
0x00,0x00,0x00,0x00,0x00,0x00,0x3F,0x40,0x40,0x40,0x40,0x40,0x40,0x78,0x00,0x00,

/*-- 文字：　八 --*/
/*-- Fixedsys12；　此字体下对应的点阵为:宽×高 = 16×16　 --*/
0x00,0x00,0x00,0x00,0x00,0xFC,0x00,0x00,0x00,0x7E,0x80,0x00,0x00,0x00,0x00,0x00,
0x00,0x80,0x60,0x18,0x07,0x00,0x00,0x00,0x00,0x00,0x03,0x0C,0x30,0x40,0x80,0x00,

/*-- 文字：　九 --*/
/*-- Fixedsys12；　此字体下对应的点阵为:宽×高 = 16×16　 --*/
0x00,0x10,0x10,0x10,0x10,0xFF,0x10,0x10,0x10,0x10,0xF0,0x00,0x00,0x00,0x00,0x00,
0x80,0x40,0x20,0x18,0x07,0x00,0x00,0x00,0x00,0x00,0x3F,0x40,0x40,0x40,0x78,0x00,

/*-- 文字：　十 --*/
/*-- Fixedsys12；　此字体下对应的点阵为:宽×高 = 16×16　 --*/

```
0x40,0x40,0x40,0x40,0x40,0x40,0x40,0xFF,0x40,0x40,0x40,0x40,0x40,0x40,0x40,0x00,
0x00,0x00,0x00,0x00,0x00,0x00,0x00,0xFF,0x00,0x00,0x00,0x00,0x00,0x00,0x00,0x00,

/*-- 文字:  正 --*/
/*-- Fixedsys12;  此字体下对应的点阵为:宽×高＝16×16   --*/
0x00,0x02,0x02,0xC2,0x02,0x02,0x02,0xFE,0x82,0x82,0x82,0x82,0x82,0x02,0x00,0x00,
0x40,0x40,0x40,0x7F,0x40,0x40,0x40,0x7F,0x40,0x40,0x40,0x40,0x40,0x40,0x40,0x00,

/*-- 文字:  腊 --*/
/*-- Fixedsys12;  此字体下对应的点阵为:宽×高＝16×16   --*/
0x00,0xFE,0x22,0x22,0xFE,0x40,0x48,0x48,0x7F,0x48,0x48,0x48,0x7F,0x48,0x48,0x00,
0x80,0x7F,0x02,0x82,0xFF,0x00,0x00,0xFF,0x49,0x49,0x49,0x49,0x49,0xFF,0x00,0x00,

/*-- 文字:  闰 --*/
/*-- Fixedsys12;  此字体下对应的点阵为:宽×高＝16×16   --*/
0x00,0xF8,0x01,0x26,0x20,0x20,0x22,0xE2,0x22,0x22,0x22,0x22,0x02,0xFE,0x00,0x00,
0x00,0xFF,0x00,0x10,0x11,0x11,0x11,0x1F,0x11,0x11,0x11,0x50,0x80,0x7F,0x00,0x00,

/*-- 文字:  廿 --*/
/*-- Fixedsys12;  此字体下对应的点阵为:宽×高＝16×16   --*/
0x20,0x20,0x20,0x20,0xFF,0x20,0x20,0x20,0x20,0x20,0x20,0xFF,0x20,0x20,0x20,0x00,
0x00,0x00,0x00,0x00,0xFF,0x40,0x40,0x40,0x40,0x40,0x40,0xFF,0x00,0x00,0x00,0x00,

};
//**************** 温度符号数据表 ****************************************

const uchar code sheshidu[][32] =
{
/*-- 文字:  ℃ --*/
/*-- Fixedsys12;  此字体下对应的点阵为:宽×高＝16×16   --*/
0x06,0x09,0x09,0xE6,0xF8,0x0C,0x04,0x02,0x02,0x02,0x02,0x02,0x04,0x1E,0x00,0x00,
0x00,0x00,0x00,0x07,0x1F,0x30,0x20,0x40,0x40,0x40,0x40,0x40,0x20,0x10,0x00,0x00,
};


#endif//
//
/*********************** 中西文字模表结束 ****************************/
```

第 8 章

课程设计一:彩灯控制器

8.1 系统功能

用 LED 发光二极管作为模拟彩灯,用单片机作为控制器,用四个按键控制 LED 小灯的工作状态,可使 LED 小灯实现轮流点亮、逐点点亮、间隔闪亮、暂停功能。实际应用时如要控制交流电大功率彩灯,可在端口加接继电器或可控硅接口电路。本设计可应用在广告彩灯控制器、舞台灯光控制器等领域。

8.2 设计方案

单片机采用 40 脚的 89C52 标准双列直插系列,有 4 个标准输入/输出端口共 32 位控制端口。本次设计采用并行口低电平(吸电流)直接驱动 LED 发光管发光形式,选择了 P1 口的 8 个端口进行模拟 LED 小灯控制,如要多些小灯单元可再将 P2 口、P0 口及其他空余端口用 LED 小灯驱动控制。因系统功能要求能控制灯亮的

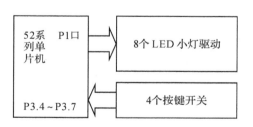

图 8.1 模拟彩灯控制器系统设计

方式,在 P3.4~P3.7 端口接了 4 个按键小开关,每个小开关可控制一种亮灯方式。在端口较紧张的情况下,LED 小灯驱动也可用串入/并出移位寄存器(如 74HC595),单片机用并行移位方式进行驱动。控制按键也可以用一个,用循环控制实现灯亮功能的转换。图 8.1 为采用 4 个按键开关控制的 8 个模拟 LED 小灯控制器的设计框图。

8.3　系统硬件仿真电路

图 8.2 为彩灯控制器的 Proteus 硬件仿真电路图。单片机采用 89C52 系列，P1 口作 LED 发光管模拟彩灯输出控制端口，P3.4～P3.7 端口接 4 个按钮小开关，用作闪烁方式控制开关。LED 发光管设计电流约为 15 mA，限流电阻为 200Ω，单片机使用 12MHz 晶振仿真调试。

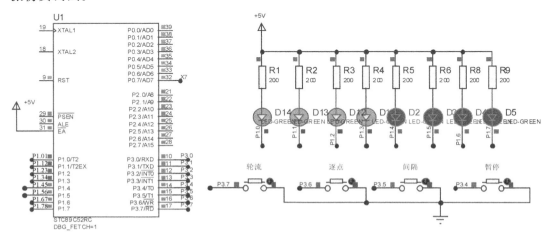

图 8.2　彩灯控制器硬件仿真电路

8.4　程序设计

（1）主体程序

通过扫描 P3.4～P3.7 端口，判断是否有按键按下，有键按下时在内存单元 20H 低四位的对应位置 1 标志，主程序通过查询标志确定应执行的闪烁方式。当 20H.0 为 1 时，发光管轮流点亮；当 20H.1 为 1 时，发光管逐点点亮；当 20H.2 为 1 时，发光管间隔闪亮；当 20H.3 为 1 时，发光管暂停闪亮。主程序对 20H 的低四位进行位值判定后，转入相应的闪烁控制程序。在上电初始化时对 20H 的最低位置 1，系统默认进入轮流点亮方式。

（2）键扫描子程序

因按键较少，采用直接端口扫描键开关，用软件延时消抖确认后对 20H 内存单元相应的位置 1 并把其余位清零。

（3）闪烁控制程序

闪烁控制程序用来控制 P1 口的发光管发光变化方式，其中执行"轮流点亮"功能程序时的 P1 口输出值变化为"11111110—延时—11111101—延时—11111011—延时—

11110111—延时—11101111—延时—11011111—延时—10111111—延时—01111111—延时—结束转主程序"。

执行"逐点点亮"功能程序时的 P1 口输出变化为"11111110—延时—11111100—延时—11111000—延 时—11110000—延 时—11100000—延 时—11000000—延 时—10000000—延时—00000000—延时—结束转主程序"。

执行"间隔闪亮"功能程序时的 P1 口输出变化为"10101010—延时—01010101—延时—结束转主程序"。

执行"暂停功能"时，对 LED 小灯端口不进行操作，保持原数据状态。

（4）延时子程序

延时子程序有 10ms 和 0.5s 等，用作键扫描消抖及 LED 发光管闪亮延时，LED 发光管闪烁的快慢可由延时程序的初值进行改变。

（5）主程序流程图

彩灯控制器设计主程序流程如图 8.3 所示。

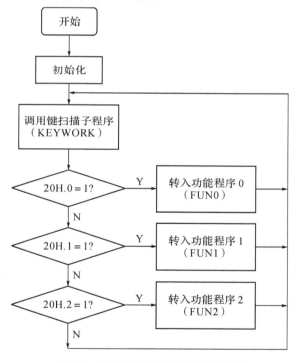

图 8.3　彩灯控制器设计主程序流程

8.5　软件调试与运行结果

电路设计好后可先写一段测试程序（将 P1 口赋♯00H），经编译后装入单片机运行，看 8 个小灯是否都会亮。按键电路可观察小开关的接地端是否为绿信号标志（小方块状),与单片机相连的端是否为红信号标志(小方块状),当鼠标按下小开关时,观察与单片机相连的端是否从红色变为绿色,否则需要检查小开关的连线情况。

电路正常后进行程序的编写,并将程序进行编译,单片机装载程序后进行运行测试,如功能不能实现再进行程序的修改。本控制器在使用中当要改变闪烁的方式时,可按下相应的功能按键,当一个完整的闪烁循环结束后转入新的闪烁方式。由于键扫描是在闪烁循环结束时进行,因此,功能开关按下的时间应较长才能被单片机读入,改进的方法是把小灯亮灯延时子程序用键扫描子程序来实现,改进后只要一按下按键即可被键扫描程序读入,程序设计调试时可以试试其控制的区别。

8.6　源程序清单

8.6.1　课程设计一参考汇编程序

```
;**********************************************;
;   课程设计一程序:彩灯控制器用 4 个按键控制      ;
;   8 个 LED 小灯的工作状态,可使 LED 小灯实现      ;
;   轮流点亮、逐点点亮、间隔闪亮功能               ;
;                  12MHz 晶振                    ;
;**********************************************;
;
LAMPOUT    EQU   P1          ;小灯输出口
KEYSW0     EQU   P3.7        ;按键 0
KEYSW1     EQU   P3.6        ;按键 1
KEYSW2     EQU   P3.5        ;按键 2
KEYSW3     EQU   P3.4        ;按键 3
;
;************;
;中断入口程序;
;************;
ORG     0000H    ;程序执行开始地址
LJMP    START    ;跳至 START 执行
;
;************;
; 初始化程序 ;
;************;
CLEAR:  MOV    20H,#00H    ;20H 单元内存清 0(闪烁标志清 0)
        SETB   00H         ;20H.0 位置 1(上电时默认执行 1 个功能)
        RET                ;子程序返回
```

```
;
;************;
;   主 程 序   ;
;************;
;
START:ACALL    CLEAR        ;调用初始化子程序
MAIN: LCALL    KEYWORK      ;调用键扫描子程序
      JB       00H,FUN0     ;20H.0 位为 1 时执行 FUN0
      JB       01H,FUN1     ;20H.1 位为 1 时执行 FUN1
      JB       02H,FUN2     ;20H.2 位为 1 时执行 FUN2
      JB       03H,MAIN     ;备用
      AJMP     MAIN         ;返回主程序 MAIN
;
;************;
;   功能程序   ;
;************;
;第 1 种闪烁功能程序
FUN0:  MOV     A,#0FEH      ;累加器赋初值
FUN00: MOV     LAMPOUT,A    ;累加器值送至 LAMPOUT 口
       LCALL   DL05S        ;延时
       JNB     ACC.7,MAIN   ;累加器最高位为 0 时转 MAIN
       RL      A            ;累加器 A 中数据循环左移 1 位
       AJMP    FUN00        ;转 FUN00 循环
;
;第 2 种闪烁功能程序
FUN1:  MOV     A,#0FEH      ;累加器赋初值
FUN11: MOV     LAMPOUT,A    ;累加器值送至 LAMPOUT 口
       LCALL   DL05S        ;延时
       JZ      MAIN         ;A 为 0 转 MAIN
       RL      A            ;累加器 A 中数据循环左移 1 位
       ANL     A,LAMPOUT    ;A 同 LAMPOUT 口值相与
       AJMP    FUN11        ;转 FUN11 循环
;
;第 3 种闪烁功能程序
FUN2:  MOV     A,#0AAH      ;累加器赋初值
       MOV     LAMPOUT,A    ;累加器值送至 LAMPOUT 口
       LCALL   DL05S        ;延时
       LCALL   DL05S        ;延时
       CPL     A            ;A 中各位取反
```

```
        MOV      LAMPOUT,A        ;累加器值送至 LAMPOUT 口
        LCALL    DL05S            ;延时
        LCALL    DL05S            ;延时
        AJMP     MAIN             ;转 MAIN
;************;
;  扫键程序  ;
;************;
;
KEYWORK:MOV      P3,♯0FFH         ;置 P3 口为输入状态
        JNB      KEYSW0,KEY0      ;读 KEYSW0 口,若为 0 转 KEY0
        JNB      KEYSW1,KEY1      ;读 KEYSW1 口,若为 0 转 KEY1
        JNB      KEYSW2,KEY2      ;读 KEYSW2 口,若为 0 转 KEY2
        JNB      KEYSW3,KEY3      ;读 KEYSW3 口,若为 0 转 KEY3
        RET                       ;子程序返回
;
;闪烁功能 0 键处理程序
KEY0:   LCALL    DL10ms           ;延时 10ms 消抖
        JB       KEYSW0,OUT0      ;KEYSW0 为 1,子程序返回(干扰)
        SETB     00H              ;20H.0 位置 1(执行闪烁功能 1 标志)
        CLR      01H              ;20H.1 位清 0
        CLR      02H              ;20H.2 位清 0
        CLR      03H              ;20H.3 位清 0
OUT0:   RET                       ;子程序返回
;
;闪烁功能 1 键处理程序
KEY1:   LCALL    DL10ms
        JB       KEYSW1,OUT1
        SETB     01H              ;20H.1 位置 1(执行闪烁功能 2 标志)
        CLR      00H
        CLR      02H
        CLR      03H
OUT1:   RET
;
;闪烁功能 2 键处理程序
KEY2:   LCALL    DL10ms
        JB       KEYSW2,OUT2
        SETB     02H              ;20H.2 位置 1(执行闪烁功能 3 标志)
        CLR      01H
        CLR      00H
```

```
              CLR     03H
OUT2：  RET
;
;闪烁功能(备用)键处理程序
KEY3：  LCALL   DL10ms
        JB      KEYSW3,OUT3
        SETB    03H              ;20H.3 位置 1(执行暂停功能标志)
        CLR     01H
        CLR     02H
        CLR     00H
OUT3：  RET
;
;***********;
;   延时程序   ;
;***********;
;约 0.5ms 延时子程序,执行一次时间为 513 μs
DL513: MOV      R2,♯0FFH
LOOP1: DJNZ     R2,LOOP1
        RET
;
;10ms 延时子程序(调用 20 次 0.5ms 延时子程序)
DL10ms:MOV      R3,♯14H
LOOP2: LCALL    DL513
        DJNZ     R3,LOOP2
        RET
;
;延时子程序,改变 R4 寄存器初值可改变闪烁的快慢(时间为 15×25ms)
DL05S: MOV      R4,♯0FH
LOOP3: LCALL    DL25ms
        DJNZ     R4,LOOP3
        RET
;
;约 25ms 延时子程序,调用扫描键子程序延时,可快速读出功能按键值
DL25ms:MOV      R5,♯0FFH
LOOP4: LCALL    KEYWORK
        DJNZ     R5,LOOP4
        RET
        END                      ;程序结束
```

8.6.2　课程设计一参考 C 程序

```
/*------------------------------------
Color LED program V7.1
MCU STC89C52RC　XAL 12MHz
Build by Gavin Hu，2010.6.10
------------------------------------ */
#include <reg51.h>
sbit KEY1 = P3^7;
sbit KEY2 = P3^6;
sbit KEY3 = P3^5;
sbit KEY4 = P3^4;
void delay_ms(unsigned int);

/*------------------------------------
  main function
------------------------------------ */
char code disp_table[][8] = {\
"\xfe\xfd\xfb\xf7\xef\xdf\xbf\x7f",\
"\xfe\xfc\xf8\xf0\xe0\xc0\x80\x00",\
"\xaa\x55\xaa\x55\xaa\x55\xaa\x55"};
void main(void)
{
char fun,i;
while(1)
    {
    if (KEY1 == 0) fun = 0;
        else if (KEY2 == 0) fun = 1;
        else if (KEY3 == 0) fun = 2;
        else if (KEY4 == 0) fun = 3;
    if (fun == 3) continue;
        else P1 = disp_table[fun][i++];
    i &= 0x07;
    delay_ms(100);
    }
}
```

```
/*------------------------------------
  Delay function
  Parameter: unsigned int dt
  Delay time = dt(ms)
  ------------------------------------ */
void delay_ms(unsigned int dt)
{
register unsigned char bt,ct;
for (;dt;dt --)
    for (ct = 2;ct;ct --)
        for (bt = 250; -- bt;);
}
```

第 9 章

课程设计二：单片机时钟

9.1 系统功能

单片机时钟要求用单片机及六位 LED 数码管显示时、分、秒，以 24 小时计时方式运行，能整点提醒（短蜂鸣，次数代表整点时间），使用按键开关可实现时分调整、秒表/时钟功能转换、省电（关闭显示）、定时设定提醒（蜂鸣器）等功能。

9.2 设计方案

为了实现 LED 显示器的数字显示，可以采用静态显示法和动态显示法，由于静态显示法需要数据锁存器等硬件，接口比较复杂，考虑时钟显示只有六位，且系统没有其他复杂的处理任务，所以采用动态扫描法实现 LED 的显示。单片机采用 89C52 系列，这样单片机就具有足够的空余硬件资源实现其他的扩充功能。单片机时钟电路系统的总体设计框架如图 9.1 所示。

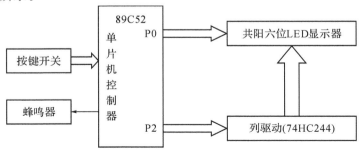

图 9.1 单片机时钟系统的总体设计框架

9.3 系统硬件仿真电路

单片机时钟硬件仿真电路见图 9.2。采用单片机最小化应用设计，采用共阳七段 LED 显示器，P0 口输出段码数据，P2.0～P2.7 口作列扫描输出，P1、P3 口串联 16 个按钮 开关后接 LED 发光管，P3.7 端口接 5V 的小蜂鸣器用于按键发音及定时提醒、整点到时 提醒等。为了提供共阳 LED 数码管的列扫描驱动电压，用 74HC244 同相驱动器作 LED 数码管的电源驱动。采用 12MHz 晶振可提高秒计时的精确性。

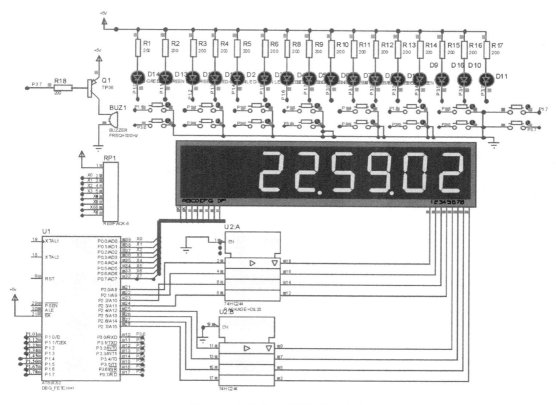

图 9.2 单片机时钟硬件仿真电路

9.4　程序设计

9.4.1　主程序

本设计中计时采用定时器 T0 中断完成，秒表使用定时器 T1 中断完成，主程序循环调用显示子程序及查键，当端口有开关按下时，转入相应功能程序。其主程序执行流程见图 9.3。

9.4.2　显示子程序

时间显示子程序每次显示 6 个连续内存单元的十进制 BCD 码数据，首地址在调用显示程序时先指定。内存中 50H～55H 为闹钟定时单元，60H～65H 为秒表计时单元，70H～75H 为时钟显示单元。由于采用七段共阳 LED 数码管动态扫描实现数据显示，显示用十进制 BCD 码数据的对应段码存放在 ROM 表中，显示时，先取出内存地址中的数据，然后查得对应的显示用段码从 P0 口输出，P2 口将对应的数码管选中供电，就能显示该地址单元的数据值。为了显示小数点及"－"、"A"等特殊字符，在开机显示班级信息和计时使用时采用不同的显示子程序。主程序流程见图 9.3。

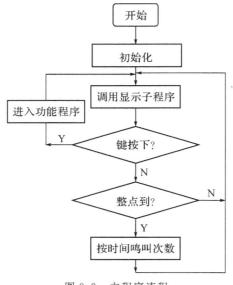

图 9.3　主程序流程

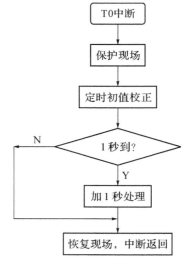

图 9.4　T0 中断计时程序流程

9.4.3 定时器 T0 中断服务程序

定时器 T0 用于时间计时。定时溢出中断周期设为 50ms，进入中断后先进行定时中断初值校正，中断累计 20 次（即 50ms×20＝1s）时对秒计数单元进行加 1 操作。时钟计数单元地址分别在 70H～71H（秒）、76H～77H（分）、78H～79H（时），最大计时值为 23 小时 59 分 59 秒。7AH 单元内存放"熄灭符"数据（♯0AH），用于时间调整时的闪烁功能。在计数单元中采用十进制 BCD 码计数，满 10 进位，T0 中断计时程序执行流程见图 9.4。

9.4.4 T1 中断服务程序

T1 中断程序用于指示时间调整单元数字的闪亮或秒表计数，在时间调整状态下，每过 0.3s 左右，将对应调整单元的显示数据换成"熄灭符"数据（♯0AH）。这样在调整时间时，对应调整单元的显示数据会间隔闪亮。在作秒表计时时，每 10ms 中断 1 次，计数单元加 1，每 100 次为 1s。秒表计数单元地址在 60H～61H（10 毫秒）、62H～63H（秒）、64H～65H（分），最大计数值为 99 分 59.99 秒。T1 中断程序流程图见图 9.5。

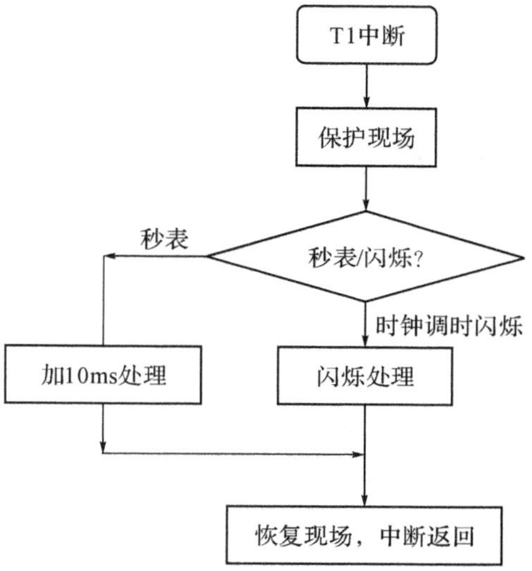

图 9.5 T1 中断程序流程

9.4.5 调时功能程序

调时功能程序的设计方法是：按下 P1.0 口按键，若按下时间小于 1s，进入省电状态（数码管不亮，时钟不停），否则进入调分状态，等待操作，此时计时器停止走动。当再按下 P1.0 按钮时，若按下时间小于 0.5s，则时间加 1min，若按下时间大于 0.5s，则进入小时调整状态，按下 P1.1 按钮时可进行减 1 调整。在小时调整状态下，当按键按下的时间大于 0.5s 时退出时间调整状态，时钟从 0s 开始计时。

9.4.6 秒表功能程序

在正常时钟状态下若按下 P1.1 口按键，则进行时钟/秒表显示功能的转换，秒表中断计时程序启动，显示首址改为 60H，LED 将显示秒表计时单元 60H～65H 中的数据。按下 P1.2 口的按键开关可实现秒表清零、秒表启动、秒表暂停功能，当再按下 P1.1 口按键时关闭 T1 秒表中断计时，显示首址又改为 70H，恢复正常时间的显示功能。

9.4.7　闹钟时间设定功能程序

在正常时钟状态下若按下 P1.3 口的按键开关，则进入设定闹时调分状态，显示首址改为 50H。LED 将显示 50H～55H 中的闹钟设定时间，显示式样为：00：00：　一，其中高 2 位代表时，低 2 位代表分，在定时闹铃时精确到分。按 P1.2 键分加 1，按 P1.0 键分减 1；若再按 P1.3 键进入时调整状态，显示式样为 00：00：　一，按 P1.2 键时加 1，按 P1.0 键时减 1，按 P1.1 键闹铃有效，显示式样变为 00：00：一0，再按 P1.1 键闹铃无效（显示式样又为 00：00：　一　）。再按 P1.3 键调整闹钟时间结束，恢复正常时间的显示。在闹铃时可按一下 P1.3 口按键使蜂鸣停止，不按，则蜂鸣器将鸣叫 1 分钟后自行中止。在设定闹钟后若要取消闹时功能，可按一下 P1.3 键，可听到一声"嘀"声表明已取消了闹铃功能。

9.5　软件调试与运行结果

在 Proteus 软件上画好电路后先要进行硬件线路的测试。先测试 LED 数码管是否会亮，方法是写一段小程序(P0 口为♯00H，P2 口为♯0FFH)，装入单片机后运行看 8 个数码管是否能显示 8 个"8"，如不会亮或部分不会亮应检查硬件连接线路；按键小开关的检查是用鼠标按下小开关看对应口的发光管是否会亮；蜂鸣器电路接在 P3.7 口，在按下 P3.7 口小开关时应能听到蜂鸣声。

单片机时钟程序的编制与调试应分段或以子程序为单位一个一个进行，最后可结合 Proteus 硬件电路调试。按照以下参考源程序，LED 显示器动态扫描的频率约为 167 次/秒，实际使用观察时完全没有闪烁现象。由于计时中断程序中加了中断延时误差处理，所以实际计时的走时精度较高，可满足一般场合的应用需要，另外上电时具有滚动显示子程序，可以方便显示制作日期等信息。

9.6　源程序清单

9.6.1　课程设计二参考汇编程序

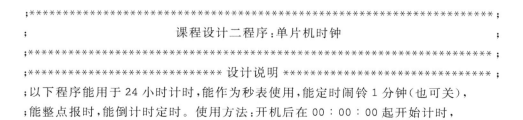

```
;**********************************************************;
;                    课程设计二程序:单片机时钟          ;
;**********************************************************;
;******************** 设计说明 ********************
;以下程序能用于 24 小时计时,能作为秒表使用,能定时闹铃 1 分钟(也可关),
;能整点报时,能倒计时定时。使用方法:开机后在 00：00：00 起开始计时,
```

```
;(1)长按 P1.0 进入调分状态:分单元闪烁,按 P1.0 加 1,按 P1.1 减 1,
;再长按 P1.0 进入时调整状态,时单元闪烁,加减调整同调分,长按 P1.0 退出调整状态。
;(2)按下 P1.1 进入秒表状态:按 P1.2 暂停,再按 P1.2 秒表清零,再按 P1.2 秒表
;又启动,按 P1.1 退出秒表回到时钟状态。
;(3)按 P1.3 进入设定闹时状态:00:00:  - ,
;可进行分设定,按 P1.2 分加 1,再按 P1.3 为时调整,00:00:  -  ,按 P1.2 时加 1,
;按 P1.1 闹铃有效,显示为 00:00:  - 0,再按 P1.1 闹铃无效(显示 00:00:  -  ,),
;按 P1.3 调闹钟结束。在闹铃时可按 P1.3 停闹,不按闹铃响 1 分钟。按 P1.4 进入倒计时
;定时模式,按 P1.5 进行分十位调整(加 1),按 P1.6 进行分个位加 1,按 P1.4 倒计时
;开始,当时间为 0 时停止倒计时,并发声提醒,倒计时过程中按 P1.4 可退回到
;正常时钟状态。定时器 T0、T1 溢出周期为 50ms,T0 为秒计数用,T1 为调整时闪烁
;及秒表定时用,P1.0、P1.1、P1.2、P1.3 为调整按钮,P0 口为字符输出口,
;P2 为扫描口,P3.7 为小喇叭口,采用共阳显示管。50H~55H 为闹钟定时单元,
;60H~65H 为秒表计时单元,70H~75H 为显示时间单元,76H~79H 为分时计时单元。
;03H 标志 = 0 时钟状态,03H = 1s;05H = 0,不闹铃,05H = 1 要闹铃;
;07H 每秒改变一次,用作间隔鸣叫。08H 为整点报时标志位,08H = 1 时为整点;
;09H 为闹铃到点标志,09H = 1 时定时闹时时间到。
;*****************************************************************;
          DISPFIRST   EQU   30H        ;显示首址存放单元
          BELL        EQU   P3.7       ;小喇叭或蜂鸣器
          CONBS       EQU   2FH        ;存放报时次数
          SONGCON     EQU   31H        ;唱歌程序计数器
          CONR2       EQU   32H        ;以下唱歌程序用寄存器
          CONR3       EQU   33H
          CONR4       EQU   34H
          CONR6       EQU   36H
          CONR7       EQU   37H
          CONR5       EQU   35H        ;以上唱歌程序寄存器
          DELAYR3     EQU   38H        ;以下延时程序用寄存器
          DELAYR5     EQU   39H        ;
          DELAYR6     EQU   3AH        ;
          DELAYR7     EQU   3BH        ;
;
;****************************************;
;              中断入口程序             ;
;****************************************;
;
          ORG  0000H                   ;程序执行开始地址
          LJMP        START            ;跳到标号 START 执行
          ORG  0003H                   ;外中断 0 中断程序入口
```

```
        RETI                          ;外中断 0 中断返回
        ORG  000BH                    ;定时器 T0 中断程序入口
        LJMP      INTT0               ;跳至 INTT0 执行
        ORG  0013H                    ;外中断 1 中断程序入口
        RETI                          ;外中断 1 中断返回
        ORG  001BH                    ;定时器 T1 中断程序入口
        LJMP      INTT1               ;跳至 INTT1 执行
        ORG  0023H                    ;串行中断程序入口地址
        RETI                          ;串行中断程序返回
;
;*********************************;
;           以下为程序开始           ;
;*********************************;
;
;整点报时功能程序
ZDBS:       MOV     A,#10             ;十位数乘 10 加上个位数为报时的次数
            MOV     B,79H
            MUL     AB
            ADD     A,78H
            MOV     CONBS,A           ;报时次数计算完成
            JZ      OUT00             ;如为午夜零点不报时
    BSLOOP: LCALL   DS20ms            ;以下按次数鸣叫
            MOV     P3,#00H
            LCALL   DL1S
            LCALL   DL1S
            MOV     P3,#0FFH
            LCALL   DL1S
            DJNZ    CONBS,BSLOOP      ;报时完成
OUT00:      CLR     08H               ;清整点报时标志
            AJMP    START1            ;返回主程序
;以下为闹钟功能时的唱歌程序
    SPPP:   ;LCALL MUSIC0             ;调用唱歌程序
            MOV     B,#10             ;闹钟叫 10 下
    BLOOP:  LCALL   DS20ms
            LCALL   DL1S              ;
            LCALL   DL1S              ;
            DJNZ    B,BLOOP
            CLR     0AH               ;清闹钟时的唱歌标志
            CLR     05H               ;清止闹标志
```

```
                AJMP    START1              ;返回主程序
;倒计时程序进入程序
DJS:            LCALL   DS20ms
                JB      P1.4,START1
   WAITH111: JNB      P1.4,WAITH111      ;等待键释放
                LJMP    DJSST
```

;**************************************;
;   主程序开始   ;
;**************************************;
;

```
        START:  MOV     SP,#80H             ;堆栈在 80H 以上
                LCALL   ST                  ;上电显示年月日及班级学号
                LCALL   STFUN0              ;流水灯
                LCALL   STMEN               ;时钟程序初始化子程序
                SETB    EA                  ;总中断开放
                SETB    ET0                 ;允许 T0 中断
                SETB    TR0                 ;开启 T0 定时器
                MOV     R4,#14H             ;1s 定时用计数值(50ms×20)
                MOV     DISPFIRST,#70H      ;显示单元为 70H～75H
                LCALL   MUSIC0              ;唱歌测试程序
;以下为主程序循环
        START1: LCALL   DISPLAY             ;调用显示子程序
                JNB     P1.0,SETMM1         ;P1.0 口为 0 时转时间调整程序
                JNB     P1.1,FUNSS          ;秒表功能,P1.1 按键调时作减 1 功能
                JNB     P1.2,FUNPT          ;秒表 STOP,PUSE,CLR
                JNB     P1.3,TSFUN          ;定时闹铃设定
                JNB     P1.4,DJS            ;倒计时功能
                JB      08H,ZDBS            ;08H 为 1,整点到,进行整点报时
                JB      0AH,SPPP            ;0AH 为 1 时,闹铃时间到,进行提醒
                AJMP    START1              ;P1.0 口为 1 时跳回 START1
;
        FUNPT:  LJMP    FUNPTT
;以下为闹铃时间设定程序,按 P1.3 进入设定
        TSFUN:  LCALL   DS20ms
                JB      P1.3,START1         ;
   WAIT113:  JNB      P1.3,WAIT113        ;等待键释放
                JB      05H,CLOSESP         ;闹铃已开的话,关闹铃
                MOV     DISPFIRST,#50H      ;显示 50H～55H 闹钟定时单元
                MOV     50H,#0CH            ;"－"闹铃设定时显示格式 00：00：－
```

```
                MOV     51H,#0AH            ;"黑"
;
        DSWAIT: SETB    EA
                LCALL   DISPLAY
                JNB     P1.2,DSFINC         ;分加 1
                JNB     P1.0,DSDEC          ;分减 1
                JNB     P1.3,DSSFU          ;进入时调整
                AJMP    DSWAIT
;
      CLOSESP: CLR      05H                 ;关闹铃标志
                CLR     BELL
                AJMP    START1
       DSSFU:  LCALL   DS20ms              ;消抖
                JB      P1.3, DSWAIT
                LJMP    DSSFUNN             ;进入时调整
;
      SETMM1:  LJMP    SETMM               ;转到时间调整程序 SETMM
;
   DSFINC:     LCALL   DS20ms              ;消抖
                JB      P1.2, DSWAIT
  DSWAIT12:    LCALL   DISPLAY             ;等键释放
                JNB     P1.2, DSWAIT12
                CLR     EA
                MOV     R0,#53H
                LCALL   ADD1               ;闹铃设定分加 1
                MOV     A,R3               ;分数据放入 A

                CLR     C                  ;清进位标志
                CJNE    A,#60H,ADDHH22
      ADDHH22: JC      DSWAIT             ;小于 60 分时返回
                ACALL   CLR0               ;大于或等于 60 分时分计时单元清 0
                AJMP    DSWAIT
   DSDEC :     LCALL   DS20ms              ;消抖
                JB      P1.0, DSWAIT
  DSWAITEE:    LCALL   DISPLAY             ;等键释放
                JNB     P1.0, DSWAITEE
                CLR     EA
                MOV     R0,#53H
                LCALL   sub1               ;闹铃设定分减 1
```

```
                    LJMP    DSWAIT
;以下为秒表功能/时钟转换程序
;按下 P1.1 可进行功能转换
        FUNSS: LCALL    DS20ms
                    JB      P1.1,START11
        WAIT11:  JNB    P1.1,WAIT11
                    CPL     03H
                    JNB     03H,TIMFUN
                    MOV     DISPFIRST,♯60H      ;显示秒表数据单元
                    MOV     60H,♯00H
                    MOV     61H,♯00H
                    MOV     62H,♯00H
                    MOV     63H,♯00H
                    MOV     64H,♯00H
                    MOV     65H,♯00H
                    MOV     TL1,♯0F0H           ;10ms 定时初值
                    MOV     TH1,♯0D8H           ;10ms 定时初值
                    SETB    TR1
                    SETB    ET1
        START11: LJMP    START1
        TIMFUN: MOV     DISPFIRST,♯70H      ;显示时钟数据单元
                    CLR     ET1
                    CLR     TR1
        START12: LJMP    START1
;以下为秒表暂停\清零功能程序
;按下 P1.2 暂停或清 0,按下 P1.1 退出秒表回到时钟计时
        FUNPTT: LCALL    DS20ms
                    JB      P1.2,START12
        WAIT22:  JNB    P1.2,WAIT21
                    CLR     ET1
                    CLR     TR1
        WAIT33:  JNB    P1.1,FUNSS
                    JB      P1.2,WAIT31
                    LCALL   DS20ms
                    JB      P1.2,WAIT33
        WAIT66:  JNB    P1.2,WAIT61
                    MOV     60H,♯00H
                    MOV     61H,♯00H
                    MOV     62H,♯00H
```

```
              MOV      63H,#00H
              MOV      64H,#00H
              MOV      65H,#00H
    WAIT44:   JNB      P1.1,FUNSS
              JB       P1.2,WAIT41
              LCALL    DS20ms
              JB       P1.2,WAIT44
    WAIT55:   JNB      P1.2,WAIT51
              SETB     ET1
              SETB     TR1
              AJMP     START1
;以下为键等待释放时显示不会熄灭程序
    WAIT21:   LCALL    DISPLAY
              AJMP     WAIT22
    WAIT31:   LCALL    DISPLAY
              AJMP     WAIT33
    WAIT41:   LCALL    DISPLAY
              AJMP     WAIT44
    WAIT51:   LCALL    DISPLAY
              AJMP     WAIT55
    WAIT61:   LCALL    DISPLAY
              AJMP     WAIT66
;
;
;**************************************;
;               1 秒计时程序               ;
;**************************************;
;T0 中断服务程序
        INTT0:   PUSH     ACC              ;累加器入栈保护
                 PUSH     PSW              ;状态字入栈保护
                 CLR      ET0              ;关 T0 中断允许
                 CLR      TR0              ;关闭定时器 T0
                 MOV      A,#0B7H          ;中断响应时间同步修正
                 ADD      A,TL0            ;低 8 位初值修正
                 MOV      TL0,A            ;重装初值(低 8 位修正值)
                 MOV      A,#3CH           ;高 8 位初值修正
                 ADDC     A,TH0            ;
                 MOV      TH0,A            ;重装初值(高 8 位修正值)
                 SETB     TR0              ;开启定时器 T0
```

```
            SETB    P3.6
            SETB    P3.5
            DJNZ    R4,OUTT0        ;20 次中断未到中断退出
    ADDSS:  MOV     R4,#14H         ;20 次中断到(1s)重赋初值
            CLR     P3.6            ;
            CLR     P3.5
            CPL     07H             ;闹铃时间隔鸣叫用
            MOV     R0,#71H         ;指向秒计时单元(71H~72H)
            ACALL   ADD1            ;调用加 1 程序(加 1s 操作)
            MOV     A,R3            ;秒数据放入 A(R3 为 2 位十进制数组合)
            CLR     C               ;清进位标志
            CJNE    A,#60H,ADDMM    ;
    ADDMM:  JC      OUTT0           ;小于 60s 时中断退出
            ACALL   CLR0            ;大于或等于 60s 时对秒计时单元清 0
            MOV     R0,#77H         ;指向分计时单元(76H~77H)
            ACALL   ADD1            ;分计时单元加 1 分钟
            MOV     A,R3            ;分数据放入 A
            CLR     C               ;清进位标志
            CJNE    A,#60H,ADDHH    ;
    ADDHH:  JC      OUTT0           ;小于 60min 时中断退出
            ACALL   CLR0            ;大于或等于 60min 时分计时单元清 0
            LCALL   DS20ms          ;正点报时
            SETB    08H
            MOV     R0,#79H         ;指向小时计时单元(78H~79H)
            ACALL   ADD1            ;小时计时单元加 1 小时
            MOV     A,R3            ;时数据放入 A
            CLR     C               ;清进位标志
            CJNE    A,#24H,HOUR     ;
    HOUR:   JC      OUTT0           ;小于 24 小时中断退出
            ACALL   CLR0            ;大于或等于 24 小时,计时单元清 0
    OUTT0:  MOV     72H,76H         ;中断退出时将分、时计时单元数据移入
            MOV     73H,77H         ;对应显示单元
            MOV     74H,78H         ;
            MOV     75H,79H         ;
            LCALL   BAOJ
            POP     PSW             ;恢复状态字(出栈)
            POP     ACC             ;恢复累加器
            SETB    ET0             ;开放 T0 中断
            RETI                    ;中断返回
```

```
;
;**********************************;
;          闪动调时程序\秒表功能程序          ;
;**********************************;
;T1 中断服务程序,用作时间调整时调整单元闪烁指示或秒表计时
INTT1:      PUSH    ACC                 ;中断现场保护
            PUSH    PSW                 ;
            JB      09H,SPCC
            JB      06H,DJSFUN
            JB      03H, MMFUN          ;03H = 1 时秒表
            MOV     TL1, ♯0B0H          ;装定时器 T1 定时初值
            MOV     TH1, ♯3CH           ;
            DJNZ    R2,INTT1OUT         ;0.3s 未到退出中断(50ms 中断 6 次)
            MOV     R2,♯06H             ;重装 0.3s 定时用初值
            CPL     02H                 ;0.3s 定时到对闪烁标志取反
            JB      02H,FLASH1          ;02H 位为 1 时显示单元"熄灭"
            MOV     72H,76H             ;02H 位为 0 时正常显示
            MOV     73H,77H             ;
            MOV     74H,78H             ;
            MOV     75H,79H             ;
INTT1OUT:   POP     PSW                 ;恢复现场
            POP     ACC                 ;
            RETI                        ;中断退出
FLASH1:     JB      01H,FLASH2          ;01H 位为 1 时,转小时熄灭控制
            MOV     72H,7AH             ;01H 位为 0 时,"熄灭符"数据放入分
            MOV     73H,7AH             ;显示单元(72H~73H),将不显示分数据
            MOV     74H,78H             ;
            MOV     75H,79H             ;
            AJMP    INTT1OUT            ;转中断退出
FLASH2:     MOV     72H,76H             ;01H 位为 1 时,"熄灭符"数据放入小时
            MOV     73H,77H             ;显示单元(74H~75H),小时数据将不显示
            MOV     74H,7AH             ;
            MOV     75H,7AH             ;
            AJMP    INTT1OUT            ;转中断退出
;
SPCC:       INC     SONGCON             ;中断服务,中断计数器加 1
            MOV     TH1,♯0D8H
            MOV     TL1,♯0EFH           ;10ms 定时初值
            AJMP    INTT1OUT            ;
```

```
DJSFUN:      LJMP     DJSS
MMFUN :      CLR      TR1
             MOV      A,#0F7H          ;中断响应时间同步修正,重装初值(10ms)
             ADD      A,TL1            ;低 8 位初值修正
             MOV      TL1,A            ;重装初值(低 8 位修正值)
             MOV      A,#0D8H          ;高 8 位初值修正
             ADDC     A,TH1            ;
             MOV      TH1,A            ;重装初值(高 8 位修正值)
             SETB     TR1              ;开启定时器 T0
             MOV      R0,#61H          ;指向秒计时单元(71H~72H)
             ACALL    ADD1             ;调用加 1 程序(加 1s 操作)
             CLR      C                ;
             MOV      A,R3             ;
             JZ       FSS1             ;加 1 后为 00,C=0
             AJMP     OUTT01           ;加 1 后不为 00,C=1
FSS1:        ACALL    CLR0             ;大于或等于 60s 时对秒计时单元清 0
             MOV      R0,#63H          ;指向分计时单元(76H~77H)
             ACALL    ADD1             ;分计时单元加 1 分钟
             MOV      A,R3             ;分数据放入 A
             CLR      C                ;清进位标志
             CJNE     A,#60H,ADDHH1    ;
ADDHH1:JC     OUTT01           ;小于 60min 时中断退出
             LCALL    CLR0             ;大于或等于 60min 时,计时单元清 0
             MOV      R0,#65H          ;指向小时计时单元(78H~79H)
             ACALL    ADD1             ;小时计时单元加 1 小时

OUTT01:
             POP      PSW              ;恢复状态字(出栈)
             POP      ACC              ;恢复累加器
             RETI                      ;中断返回
;*****************************************;
;              加 1 子程序                ;
;*****************************************;
;
;
ADD1:   MOV      A,@R0            ;取当前计时单元数据到 A
             DEC      R0               ;指向前一地址
             SWAP     A                ;A 中数据高四位与低四位交换
             ORL      A,@R0            ;前一地址中数据放入 A 中低四位
```

```
        ADD     A,#01H          ;A 加 1 操作
        DA      A               ;十进制调整
        MOV     R3,A            ;移入 R3 寄存器
        ANL     A,#0FH          ;高四位变 0
        MOV     @R0,A           ;放回前一地址单元
        MOV     A,R3            ;取回 R3 中暂存数据
        INC     R0              ;指向当前地址单元
        SWAP    A               ;A 中数据高四位与低四位交换
        ANL     A,#0FH          ;高四位变 0
        MOV     @R0,A           ;数据放入当前地址单元中
        RET                     ;子程序返回
;
;*********************************************;
;               分减 1 子程序                 ;
;*********************************************;
;
SUB1:   MOV     A,@R0           ;取当前计时单元数据到 A
        DEC     R0              ;指向前一地址
        SWAP    A               ;A 中数据高四位与低四位交换
        ORL     A,@R0           ;前一地址中数据放入 A 中低四位
        JZ      SUB11
        DEC     A               ;A 减 1 操作
SUB111: MOV     R3,A            ;移入 R3 寄存器
        ANL     A,#0FH          ;高四位变 0
        CLR     C               ;清进位标志
        SUBB    A,#0AH
SUB1111:JC      SUB1110
        MOV     @R0,#09H        ;大于等于 0AH,为 9
SUB110: MOV     A,R3            ;取回 R3 中暂存数据
        INC     R0              ;指向当前地址单元
        SWAP    A               ;A 中数据高四位与低四位交换
        ANL     A,#0FH          ;高四位变 0
        MOV     @R0,A           ;数据放入当前地址单元中
        RET                     ;子程序返回
;
SUB11:  MOV     A,#59H
        AJMP    SUB111
SUB1110:MOV     A,R3            ;移入 R3 寄存器
        ANL     A,#0FH          ;高四位变 0
```

```
        MOV     @R0,A
        AJMP    SUB110
```

;**;
;　　　　　　　　　时减 1 子程序　　　　　　　　　　；
;**;
;

```
    SUBB1：  MOV     A,@R0               ;取当前计时单元数据到 A
            DEC     R0                  ;指向前一地址
            SWAP    A                   ;A 中数据高四位与低四位交换
            ORL     A,@R0               ;前一地址中数据放入 A 中低四位
            JZ      SUBB11              ;00 减 1 为 23（小时）
            DEC     A                   ;A 减 1 操作
    SUBB111： MOV     R3,A               ;移入 R3 寄存器
            ANL     A,♯0FH              ;高四位变 0
            CLR     C                   ;清进位标志
            SUBB    A,♯0AH              ;时个位大于 9,为 9
    SUBB1111：JC     SUBB1110            ;
            MOV     @R0,♯09H            ;大于等于 0AH,为 9
    SUBB110： MOV     A,R3               ;取回 R3 中暂存数据
            INC     R0                  ;指向当前地址单元
            SWAP    A                   ;A 中数据高四位与低四位交换
            ANL     A,♯0FH              ;高四位变 0
            MOV     @R0,A               ;时十位数数据放入
            RET                         ;子程序返回
;

    SUBB11：MOV      A,♯23H
           AJMP     SUBB111
    SUBB1110：MOV    A,R3               ;时个位小于 0A 不处理
            ANL      A,♯0FH             ;高四位变 0
            MOV      @R0,A              ;个位移入
            AJMP     SUBB110
```

;**
;;　　　　　　　　　清零程序　　　　　　　　　　;;
;**
;对计时单元复零用

```
        CLR0：CLR   A                   ;清累加器
             MOV    @R0,A               ;清当前地址单元
             DEC    R0                  ;指向前一地址
             MOV    @R0,A               ;前一地址单元清 0
```

```
            RET                            ;子程序返回
;
;***************************************;
;                时钟时间调整程序            ;
;***************************************;
;当调时按键按下时进入此程序
       SETMM:    cLR    ET0                ;关定时器 T0 中断
                 CLR    TR0                ;关闭定时器 T0
                 LCALL  DL1S               ;调用 1s 延时程序
                 LCALL  DS20ms             ;消抖
                 JB     P1.0,CLOSEDIS      ;键按下时间小于 1s,关闭显示(省电)
                 MOV    R2,#06H            ;进入调时状态,赋闪烁定时初值
                 MOV    70H,#00H           ;调时时秒单元为 00s
                 MOV    71H,#00H
                 SETB   ET1                ;允许 T1 中断
                 SETB   TR1                ;开启定时器 T1
        SET2:    JNB    P1.0,SET1          ;P1.0 口为 0(键未释放),等待
                 SETB   00H                ;键释放,分调整闪烁标志置 1
        SET4:    JB     P1.0,SET3          ;等待键按下
                 LCALL  DL05S              ;有键按下,延时 0.5s
                 LCALL  DS20ms             ;消抖
                 JNB    P1.0,SETHH         ;按下时间大于 0.5s 转调小时状态
                 MOV    R0,#77H            ;按下时间小于 0.5s 加 1 分钟操作
                 LCALL  ADD1               ;调用加 1 子程序
                 MOV    A,R3               ;取调整单元数据
                 CLR    C                  ;清进位标志
                 CJNE   A,#60H,HHH         ;调整单元数据与 60 比较
        HHH: JC  SET4                      ;调整单元数据小于 60 转 SET4 循环
                 LCALL  CLR0               ;调整单元数据大于或等于 60 时清 0
                 CLR    C                  ;清进位标志
                 AJMP   SET4               ;跳转到 SET4 循环
     CLOSEDIS:   SETB   ET0                ;省电(LED 不显示)状态,开 T0 中断
                 SETB   TR0                ;开启 T0 定时器(开时钟)
        CLOSE:   JB     P1.0,CLOSE         ;无按键按下,等待
                 LCALL  DS20ms             ;消抖
                 JB     P1.0,CLOSE         ;是干扰,返回 CLOSE 等待
       WAITH:    JNB    P1.0,WAITH         ;等待键释放
                 LJMP   START1             ;返回主程序(LED 数据显示亮)
       SETHH:    CLR    00H                ;分闪烁标志清除(进入调小时状态)
```

```
              SETB    01H              ;小时调整标志置1
    SETHH1：  JNB     P1.0,SET5        ;等待键释放
      SET6：  JB      P1.0,SET7        ;等待按键按下
              LCALL   DL05S            ;有键按下延时0.5s
              LCALL   DS20ms           ;消抖
              JNB     P1.0,SETOUT      ;按下时间大于0.5s退出时间调整
              MOV     R0,#79H          ;按下时间小于0.5s加1小时操作
              LCALL   ADD1             ;调加1子程序
              MOV     A,R3             ;
              CLR     C                ;
              CJNE    A,#24H,HOUU      ;计时单元数据与24比较
      HOUU：  JC      SET6             ;小于24转SET6循环
              LCALL   CLR0             ;大于或等于24时清0操作
              AJMP    SET6             ;跳转到SET6循环
    SETOUT：  JNB     P1.0,SETOUT1     ;调时退出程序,等待键释放
              LCALL   DS20ms           ;消抖
              JNB     P1.0,SETOUT      ;是抖动,返回SETOUT再等待
              CLR     01H              ;清调小时标志
              CLR     00H              ;清调分标志
              CLR     02H              ;清闪烁标志
              CLR     TR1              ;关闭定时器T1
              CLR     ET1              ;关定时器T1中断
              SETB    TR0              ;开启定时器T0
              SETB    ET0              ;开定时器T0中断(计时开始)
              LJMP    START1           ;跳回主程序
      SET1：  LCALL   DISPLAY          ;键释放等待时调用显示程序(调分)
              AJMP    SET2             ;防止键按下时无时钟显示
      SET3：  LCALL   DISPLAY          ;等待调分按键时时钟显示用
              JNB     P1.1,FUNSUB      ;减1分操作
              AJMP    SET4             ;调分等待
      SET5：  LCALL   DISPLAY          ;键释放等待时调用显示程序(调小时)
              AJMP    SETHH1           ;防止键按下时无时钟显示
      SET7：  LCALL   DISPLAY          ;等待调小时按键时时钟显示用
              JNB     P1.1,FUNSUBB     ;小时减1操作
              AJMP    SET6             ;调时等待
   SETOUT1：  LCALL   DISPLAY          ;退出时钟调整时键释放等待
              AJMP    SETOUT           ;防止键按下时无时钟显示
;FUNSUB,分减1程序
    FUNSUB：  LCALL   DS20ms           ;消抖
```

```
            JB      P1.1,SET41          ;干扰,返回调分等待
  FUNSUB1:  JNB     P1.1,FUNSUB15       ;等待键放开
            MOV     R0,#77H             ;
            LCALL   SUB1                ;分减 1 程序
            LJMP    SET4                ;返回调分等待
;
    SET41:  LJMP    SET4                ;
;FUNSUBB,时减 1 程序
  FUNSUBB:  LCALL   DS20ms              ;消抖
            JB      P1.1,SET61          ;干扰,返回调时等待
  FUNSUBB1:JNB      P1.1,FUNSUBB1       ;等待键放开
            MOV     R0,#79H             ;
            LCALL   SUBB1               ;时减 1 程序
            LJMP    SET6                ;返回调时等待
;
        SET61:  LJMP    SET6
;*******************************************;
;                  显示程序                 ;
;*******************************************;
;显示数据在 70H~75H 单元内,用六位 LED 共阳数码管显示,P0 口输出段码数据,
;P2 口作扫描控制,每个 LED 数码管亮 1ms 时间再逐位循环
        DISPLAY: MOV    R1,DISPFIRST     ;指向显示数据首址
                 MOV    R5,#80H          ;扫描控制字初值
        PLAY:    MOV    A,R5             ;扫描字放入 A
                 MOV    P2,A             ;从 P2 口输出
                 MOV    A,@R1            ;取显示数据到 A
                 MOV    DPTR,#TAB        ;取段码表地址
                 MOVC   A,@A+DPTR        ;查显示数据对应段码
                 MOV    P0,A             ;段码放入 P1 口
                 MOV    A,R5             ;
                 JNB    ACC.5,LOOP5      ;小数点处理
                 CLR    P0.7             ;
        LOOP5:   JNB    ACC.3,LOOP6      ;小数点处理
                 CLR    P0.7             ;
        LOOP6:   LCALL  DL1ms            ;显示 1ms
                 INC    R1               ;指向下一地址
                 MOV    A,R5             ;扫描控制字放入 A
                 JB     ACC.2,ENDOUT     ;ACC.5 = 0 时一次显示结束
                 RR     A                ;A 中数据循环左移
```

```
            MOV     R5,A                    ;放回 R5 内
            MOV     P0,♯0FFH
            AJMP    PLAY                    ;跳回 PLAY 循环
    ENDOUT：MOV     P2,♯00H                 ;一次显示结束,P2 口复位
            MOV     P0,♯0FFH                ;P0 口复位
            RET                             ;子程序返回
TAB：DB 0C0H,0F9H,0A4H,0B0H,99H,92H,82H,0F8H,80H,90H,0FFH,88H,0BFH
;共阳段码表    "0""1""2" "3""4""5""6""7" "8""9""不亮""A"" - "
;
;*********************************** ;
;     SDISPLAY,上电滚动显示子程序              ;
;*********************************** ;
;不带小数点显示,有"A"" - "显示功能
    SDISPLAY：MOV   R1,DISPFIRST
              MOV   R5,♯80H               ;扫描控制字初值
    SPLAY：  MOV    A,R5                   ;扫描字放入 A
             MOV    P2,A                   ;从 P2 口输出
             MOV    A,@R1                  ;取显示数据到 A
             MOV    DPTR,♯TABS             ;取段码表地址
             MOVC   A,@A + DPTR            ;查显示数据对应段码
             MOV    P0,A                   ;段码放入 P1 口
             MOV    A,R5                   ;
             LCALL  DL1ms                  ;显示 1ms
             INC    R1                     ;指向下一地址
             MOV    A,R5                   ;扫描控制字放入 A
             JB     ACC.2,ENDOUTS          ;ACC.5 = 0 时一次显示结束
             RR     A                      ;A 中数据循环左移
             MOV    R5,A                   ;放回 R5 内
             AJMP   SPLAY                  ;跳回 PLAY 循环
    ENDOUTS：MOV    P2,♯00H                ;一次显示结束,P2 口复位
             MOV    P0,♯0FFH               ;P0 口复位
             RET                           ;子程序返回
TABS：DB 0C0H,0F9H,0A4H,0B0H,99H,92H,82H,0F8H,80H,90H,0FFH,0C6H,0BFH,88H
;显示数  "0    1    2    3    4  5  6  7  8    9    不亮  C    -    A"
;内存数  "0    1    2    3    4  5  6  7  8    9    0AH  0BH 0CH  0DH "
;STAB 表,启动时显示 2006 年 12 月 23 日、C04 - 2 - 28(学号)用
STAB：DB 0AH,0AH,0AH,0AH,0AH,0AH,08H,02H,0CH,02H,0CH,04H,00H,0BH,0AH,0AH
DB 03H,02H,0CH,02H,01H,0CH,06H,00H,00H,02H,0AH,0AH,0AH,0AH,0AH,0AH
;注:0A 不亮,0B 显示"A",0C 显示" - "
```

```
;
;**********************************************************;
;   ST,上电时显示年月班级用,采用移动显示,先右移,接着左移          ;
;**********************************************************;
        ST:     MOV    R0,#40H              ;将显示内容移入 40H－5FH 单元
                MOV    R2,#20H              ;
                MOV    R3,#00H              ;
                MOV    R4,#0FEH
                MOV    P1,R4
                CLR    A                    ;
                MOV    DPTR,#STAB           ;
        SLOOP:  MOVC   A,@A+DPTR            ;
                MOV    @R0,A                ;
                MOV    A,R3                 ;
                INC    A                    ;
                MOV    R3,A                 ;
                INC    R0                   ;
                DJNZ   R2,SLOOP             ;移入完毕
                MOV    DISPFIRST,#5AH       ;以下程序从右往左移
                MOV    R3,#1BH              ;显示 27 个单元
    SSLOOP2:    MOV    R2,#25               ;控制移动速度
    SSLOOP12:   LCALL  SDISPLAY             ;
                DJNZ   R2,SSLOOP12          ;
                MOV    A,R4
                RL     A
                MOV    R4,A
                MOV    P1,A
                DEC    DISPFIRST
                DJNZ   R3,SSLOOP2           ;
                MOV    P1,#0FFH
                MOV    DISPFIRST,#40H       ;以下程序从左往右移
    SSLOOP:     MOV    R2,#25               ;控制移动速度
    SSLOOP1:    LCALL  SDISPLAY             ;
                DJNZ   R2,SSLOOP1           ;
                MOV    A,R4
                RL     A
                MOV    R4,A
                MOV    P3,A
                INC    DISPFIRST
```

```
              MOV     A,DISPFIRST
              CJNE    A,#5AH,SSLOOP     ;
              MOV     P3,#0FFH
              RET
```

;***;
;　　　　　　　　　　延时程序　　　　　　　　　　　　　;
;***;
;

;1ms 延时程序, LED 显示程序用
```
    DL1ms:    MOV     DELAYR6,#14H
    DL1:      MOV     DELAYR7,#19H
    DL2:      DJNZ    DELAYR7,DL2
              DJNZ    DELAYR6,DL1
              RET
    DL50ms:   MOV     DELAYR5,#50
    DLms:     LCALL   DL1ms
              DJNZ    DELAYR5,DLms
              RET
```
;20ms 延时程序, 采用调用显示子程序以改善 LED 的显示闪烁现象
```
    DS20ms:   CLR     BELL
              LCALL   DISPLAY
              LCALL   DISPLAY
              LCALL   DISPLAY
              SETB    BELL
              RET
```
;延时程序, 用作按键时间的长短判断
```
    DL1S:     LCALL   DL05S
              LCALL   DL05S
              RET
    DL05S:    MOV     DELAYR3,#20H      ;8ms * 32 = 0.196s
    DL05S1:   LCALL   DISPLAY
              DJNZ    DELAYR3,DL05S1
              RET
```

;***;
;以下是闹铃时间设定程序中的时调整程序　;
;***;
```
DSSFUNN:      LCALL   DISPLAY           ;等待键释放
              JNB     P1.3,DSSFUNN
              MOV     50H,#0AH          ;时调整时显示为 00：00 :-
```

```
                    MOV      51H,＃0CH
         WAITSS:     SETB     EA
                    LCALL    DISPLAY
                    JNB      P1.2,FFFF              ;时加 1 键
                    JNB      P1.0,DDDD              ;时减 1
                    JNB      P1.3,OOOO              ;闹铃设定退出键
                    JNB      P1.1,ENA              ;闹铃设定有效或无效按键
                    AJMP     WAITSS
         OOOO:      LCALL    DS20ms                 ;消抖
                    JB       P1.3, WAITSS
         DSSFUNNM:   LCALL    DISPLAY               ;键释放等待
                    JNB      P1.3, DSSFUNNM
                    MOV      DISPFIRST,＃70H
                    LJMP     START1
         ENA:       LCALL    DS20ms                 ;消抖
                    JB       P1.1, WAITSS
         DSSFUNMMO:  LCALL    DISPLAY               ;键释放等待
                    JNB      P1.1, DSSFUNMMO
                    CPL      05H
                    JNB      05H,WAITSS11
                    MOV      50H,＃00H              ;05H＝1,闹铃开,显示为 00：00：0
                    AJMP     WAITSS
         WAITSS11:   MOV      50H,＃0aH              ;闹铃不开,显示为 00：00：-
                    AJMP     WAITSS
         FFFF:      LCALL    DS20ms                 ;消抖
                    JB       P1.2, WAITSS
         DSSFUNMM:   LCALL    DISPLAY               ;键释放等待
                    JNB      P1.2, DSSFUNMM
                    CLR      EA
                    MOV      R0,＃55H
                    LCALL    ADD1
                    MOV      A,R3                  ;

                    CLR      C                      ;
                    CJNE     A,＃24H,ADDHH33N        ;
         ADDHH33N:   JC       WAITSS                ;小于 24 点返回
                    ACALL    CLR0                  ;大于等于 24 点清零
                    AJMP     WAITSS
         DDDD:      LCALL    DS20ms                 ;消抖
```

```
                    JB      P1.0，WAITSS
        DSSFUNDD：   LCALL   DISPLAY                ;键释放等待
                    JNB     P1.0，DSSFUNDD
                    CLR     EA
                    MOV     R0，#55H
                    LCALL   SUBB1
                    LJMP    WAITSS
```

```
;********************;
;以下是闹铃判断子程序  ；
;********************;
```

```
BAOJ：      JNB     05H，BBAO            ;05H = 1,闹钟开,要比较数据
            MOV     A，79H               ;从时十位、时个位、分十位、分个位顺序比较
            CJNE    A，55H，BBAO
            MOV     A，78H
            CLR     C
BB3：        CJNE    A，54H，BBAO
            MOV     A，77H
            CLR     C
            CJNE    A，53H，BBAO
            MOV     A，76H
            CLR     C
BB2：        CJNE    A，52H，BBAO
            SETB    0AH
    ;       JNB     07H，BBAO            ;07H 在 1s 到时会取反
    ;       CLR     BELL                ;时分相同时鸣叫(1s 间隔叫)
            RET
;
        BBAO：   SETB    BELL             ;不相同或闹铃没开关蜂鸣器

            RET
```

```
;*************************************************
;倒计时调分十位数
SADD：       LCALL   DS20ms
            JB      P1.5，LOOOP
        SADDWAIT：JNB   P1.5，SADDWAIT
            INC     65H                    ;十位加 1
            MOV     A，#9
            SUBB    A，65H
            JNC     LOOOP
```

```
              MOV      65H,♯00H          ;大于 9 为 0
              AJMP     LOOOP
;倒计时调分个位数
GADD:         LCALL    DS20ms
              JB       P1.6,LOOOP
    GADDWAIT:JNB       P1.6,GADDWAIT
              INC      64H               ;十位加 1
              MOV      A,♯9
              SUBB     A,64H
              JNC      LOOOP
              MOV      64H,♯00H          ;大于 9 为 0
              AJMP     LOOOP
;倒计时程序
DJSST:
              CPL      06H
              JNB      06H,TIMFUNN
              MOV      DISPFIRST,♯60H    ;显示秒表数据单元
              MOV      60H,♯00H
              MOV      61H,♯00H
              MOV      62H,♯00H
              MOV      63H,♯00H
              MOV      64H,♯01H
              MOV      65H,♯00H
              MOV      TL1,♯0F0H         ;10ms 定时初值( )
              MOV      TH1,♯0D8H         ;10ms 定时初值
    LOOOP:    LCALL    DISPLAY           ;倒计时准备,等待键按下
              JNB      P1.5,SADD
              JNB      P1.6,GADD
              JB       P1.4,LOOOP
              LCALL    DS20ms
              JB       P1.4,LOOOP
              SETB     TR1               ;倒计时开始
              SETB     ET1
    LOOOPP:   LCALL    DISPLAY
              JNB      P1.4,LOOOPP
  START11222: LJMP     START1
    TIMFUNN: MOV      DISPFIRST,♯70H    ;显示计时数据单元
              CLR      ET1
              CLR      TR1
```

```
                    LJMP    START1
    ；
    DJSS：          CLR     TR1
                    MOV     A,♯0F7H          ；中断响应时间同步修正,重装初值(10ms)
                    ADD     A,TL1            ；低 8 位初值修正
                    MOV     TL1,A            ；重装初值(低 8 位修正值)
                    MOV     A,♯0D8H          ；高 8 位初值修正
                    ADDC    A,TH1            ；
                    MOV     TH1,A            ；重装初值(高 8 位修正值)
                    SETB    TR1              ；开启定时器 T0
                    MOV     A,61H
                    SWAP    A
                    ORL     A,60H
                    JZ      FSS111
                    SUBB    A,♯01H
                    MOV     R3,A
                    ANL     A,♯0F0H
                    SWAP    A
                    MOV     61H,A
                    MOV     A,R3
                    ANL     A,♯0FH
                    MOV     60H,A
                    CJNE    A,♯0AH,JJJ
            JJJ：   JC      OUTT011
                    MOV     60H,♯09
                    AJMP    OUTT011          ；加 1 后不为 00,C＝1
    FSS111：        MOV     60H,♯09
                    MOV     61H,♯09
                    MOV     A,63H
                    SWAP    A
                    ORL     A,62H
                    JZ                       FSS222
                    SUBB    A,♯01H
                    MOV     R3,A
                    ANL     A,♯0F0H
                    SWAP    A
                    MOV     63H,A
                    MOV     A,R3
                    ANL     A,♯0FH
```

```
                    MOV     62H,A
                    CJNE    A,#0AH,KKK
        KKK：       JC      OUTTO11
                    MOV     62H,#09
                    AJMP    OUTTO11             ;加 1 后不为 00,C = 1
        FSS222：    MOV     62H,#09
                    MOV     63H,#05             ;小于 60min 时中断退出
                    MOV     A,65H
                    SWAP    A
                    ORL     A,64H
                    JZ                          FSS333
                    SUBB    A,#01H
                    MOV     R3,A
                    ANL     A,#0F0H
                    SWAP    A
                    MOV     65H,A
                    MOV     A,R3
                    ANL     A,#0FH
                    MOV     64H,A
                    CJNE    A,#0AH,qqq
        qqq：       JC      OUTTO11
                    MOV     64H,#09
                    AJMP    OUTTO11             ;加 1 后不为 00,C = 1
        FSS333：    MOV     64H,#00
                    MOV     65H,#00
                    MOV     63H,#00
                    MOV     62H,#00
                    MOV     61H,#00
                    MOV     60H,#00
                    CLR     BELL
                    CLR     TR1
                    CLR     ET1
        OUTTO11：
                    POP     PSW                 ;恢复状态字(出栈)
                    POP     ACC                 ;恢复累加器
                    RETI                        ;中断返回
    ;
    ;开启流水灯子程序
        STFUN0：    MOV     A,#0FEH             ;
```

```
        FUN0011：   MOV     P1,A              ;
                    LCALL   DL50ms            ;
                    JNB     ACC.7,MAINEND     ;
                    RL      A                 ;
                    AJMP    FUN0011           ;
        MAINEND：   MOV     P1,♯0FFH
                    MOV     A,♯0FEH           ;
        FUN0022：   MOV     P3,A              ;
                    LCALL   DL50ms            ;
                    JNB     ACC.7,MAINEND1    ;
                    RL      A                 ;
                    AJMP    FUN0022           ;
        MAINEND1：  MOV     P3,♯0FFH
                    RET
;时钟程序开始初始化程序
STMEN：             MOV     R0,♯20H           ;清 20H～7FH 内存单元
                    MOV     R7,♯60H           ;
        CLEARDISP： MOV     @R0,♯00H          ;
                    INC     R0                ;
                    DJNZ    R7,CLEARDISP      ;
                    MOV     20H,♯00H          ;清 20H(标志用)
                    MOV     21H,♯00H          ;清 21H(标志用)
                    MOV     7AH,♯0AH          ;放入"熄灭符"数据
                    MOV     TMOD,♯11H         ;设 T0、T1 为 16 位定时器
                    MOV     TL0,♯0B0H         ;50ms 定时初值(T0 计时用)
                    MOV     TH0,♯3CH          ;50ms 定时初值
                    MOV     TL1,♯0B0H         ;50ms 定时初值(T1 闪烁定时用)
                    MOV     TH1,♯3CH          ;50ms 定时初值
                    RET
;以下为唱歌程序
MUSIC0：            MOV     TH1,♯0D8H
                    MOV     TL1,♯0EFH
                    SETB    TR1
                    SETB    09H
                    SETB    ET1
                    MOV     DPTR,♯DAT         ;表头地址送 DPTR
                    MOV     SONGCON,♯00H      ;中断计数器清 0
;
MUSIC1：
```

```
              CLR      A
              MOVC     A,@A+DPTR          ;查表取代码
              JZ       END0               ;是 00H,则结束
              CJNE     A,♯0FFH,MUSIC5
              LJMP     MUSIC3
     MUSIC5:
              MOV      CONR6,A
              INC      DPTR
              CLR      A
              MOVC     A,@A+DPTR          ;取节拍代码送 R7
              MOV      CONR7,A
              SETB     TR1                ;启动计数
     MUSIC2:
              CPL      BELL
              MOV      A,CONR6
              MOV      CONR3,A
              LCALL    DEL
              MOV      A,CONR7
              CJNE     A,SONGCON,MUSIC2   ;中断计数器(SONGCON)=R7 是否成立
              MOV      SONGCON,♯00H       ;等于,则取下一代码
              INC      DPTR
              LJMP     MUSIC1
     MUSIC3:
              CLR      TR1                ;休止 100ms
              MOV      CONR2,♯0DH
     MUSIC4:
              NOP
              MOV      CONR3,♯0FFH
              LCALL    DEL
              DJNZ     CONR2,MUSIC4
              INC      DPTR
              LJMP     MUSIC1
     END0:    CLR      ET1
              CLR      TR1
              CLR      09H
              RET
     DEL:
              NOP
     DEL3:
              MOV      CONR4,♯02H
```

```
DEL4：
                    NOP
                    DJNZ    CONR4,DEL4
                    NOP
                    DJNZ    CONR3,DEL3
                    RET
```

;以下为唱歌音乐代码表,数据为音调－节拍,FF 为休止 100ms,00H 为结束

```
DAT：
db 26h,20h,20h,20h,20h,20h,26h,10h,20h,10h,20h,80h,26h,20h,30h,20h
db 30h,20h,39h,10h,30h,10h,30h,80h,26h,20h,20h,20h,20h,20h,1ch,20h
db 20h,80h,2bh,20h,26h,20h,20h,20h,2bh,10h,26h,10h,2bh,80h,00H
;
                    END                              ;程序结束
```

9.6.2　课程设计二参考 C 程序

```c
/*——————————————————————————————
Clock program V8.1
MCU STC89C52RC   XAL 12MHz
Build by Gavin Hu, 2007.12.16
—————————————————————————————— */
# include <reg51.h>
//
# define uchar unsigned char
# define uint unsigned int
# define ulong unsigned long
sbit BUZZ = P3^7;
sbit KEY1 = P1^0;
sbit KEY2 = P1^1;
uchar hour_reg, minute_reg, second_reg;

void delay(uint);
void display(uchar * );
void time2str(uchar * );
void time_set(void);

/*——————————————————————————————
  main function
—————————————————————————————— */
```

```c
void main(void)
{
uchar dispram[9];
TMOD = 0x11;
IE = 0x82;
TR0 = 1;
while (1)
    {
    time2str(dispram);
    display(dispram);
    if (KEY1 == 0) time_set();
    }
}


/*----------------------------------------
   Time data to display string function
   Parameter : pointer of string
------------------------------------ */
void time2str(uchar * ch)
{
ch[0] = hour_reg/10;
ch[1] = hour_reg % 10;
ch[2] = 16;
ch[3] = minute_reg/10;
ch[4] = minute_reg % 10;
ch[5] = 16;
ch[6] = second_reg/10;
ch[7] = second_reg % 10;
}


/*----------------------------------------
   Set time function
------------------------------------ */
void time_set(void)
{
uchar ch[8];
uchar i,c;
TR0 = 0;
second_reg = 0;
```

```
    time2str(ch);
do {
    display(ch);
    } while (KEY1 == 0);
c = 2;
while (c)
    {
    time2str(ch);
    if (c == 2) {ch[0]| = 0x80;ch[1]| = 0x80;}
        else {ch[3]| = 0x80;ch[4]| = 0x80;}
    display(ch);
    if (KEY1 == 0)
        {
        c -- ;
        do {
            display(ch);
            } while (KEY1 == 0);
        }
    if (KEY2 == 0)
        {
        if (c == 2) hour_reg = (hour_reg + 1) % 24;
            else minute_reg = (minute_reg + 1) % 60;
        for (i = 0;i<50;i ++ ) display(ch);
        }
    }
TR0 = 1;
}


/*------------------------------------
  Delay function
  Parameter: unsigned int dt
  Delay time = dt(ms)
  ------------------------------------ */
void delay(unsigned int dt)
{
register unsigned char bt,ct;
for (;dt;dt --)
    for (ct = 2;ct;ct --)
        for (bt = 250; -- bt;);
}
```

```
/*-------------------------------------
  8 LED digital tubes display function
  Parameter: sting pointer to display
------------------------------------- */
void display(uchar * disp_ram)
{
static uchar disp_count;
unsigned char i,j;
unsigned char code table[] =
{0xc0,0xf9,0xa4,0xb0,0x99,0x92,0x82,0xf8,0x80,0x90,0x88,
0x83,0xc6,0xa1,0x86,0x8e,0xbf,0xff};
disp_count = (disp_count + 1)&0x7f;
for (i = 0;i<8;i ++ )
    {
    j = disp_ram[i];
    if (j&0x80) P0 = (disp_count>32)? table[j&0x7f]:0xff;
        else P0 = table[j];
    P2 = 0x01<<i;
    delay(1);
    P0 = 0xff;
    P2 = 0;
    }
}


/*-------------------------------------
  Time function(using T0 interrupt)
------------------------------------- */
void T0_time(void)interrupt 1
{
static uchar T0_count = 0;
TL0 = 0xb0;
TH0 = 0x3c;
T0_count ++ ;
if (T0_count> = 20)
    {
    T0_count = 0;
    second_reg ++ ;
    if (second_reg> = 60)
        {
```

```
        second_reg = 0;
        minute_reg ++ ;
        if (minute_reg> = 60)
            {
            minute_reg = 0;
            hour_reg = (hour_reg + 1) % 24;
            }
        }
    }
}
```

第 10 章

课程设计三：DS1302 实时时钟

10.1 系统功能

DS1302 实时时钟芯片能输出阳历年、月、日及星期、小时、分、秒等计时信息，可制作成实时时钟。本系统要求用八位 LED 数码管实时显示时、分、秒时间。

10.2 设计方案

按照系统设计功能的要求，确定由主控模块、时钟模块、显示模块、键盘接口模块、发声模块共 5 个模块组成，电路系统构成框图见图 10.1。主控芯片使用 89C52 系列单片机，时钟芯片使用美国 DALLAS 公司推出的一种高性能、低功耗、带 RAM 的实时时钟 DS1302，采用 DS1302 作为计时芯片，可以做到计时准确，更重要的是 DS1302 可以在很小电流的后备电源（2.5～5.5V 电源，在 2.5V 时耗电小于 300nA）下继续计时，而且 DS1302 可以编程选择多种充电电流来对后备电源进行慢速充电，可以保证后备电源基本不耗电。显示电路采用八位共阳 LED 数码管，采用查询法查键实现功能调整。

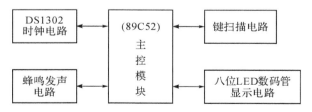

图 10.1 DS1302 实时时钟电路系统构成

10.3　系统硬件仿真电路

DS1302 实时时钟的硬件仿真电路见图 10.2。时钟芯片的晶振频率为 32.768kHz，3 个数据、时钟、片选口可不接上拉电阻；LED 数码管采用动态扫描方式显示，P0 口为段码输出口，P2 口为扫描驱动口，扫描驱动信号经 74HC244 功率放大用作 LED 点亮电源；调时按键设计了 2 个，分别接在 P3.5、P3.6 口，用于设定与加 1 调整；P3.7 口接了一个蜂鸣发声器，用于按键发声提醒用。

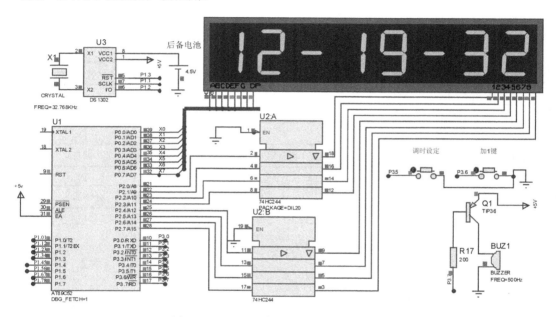

图 10.2　DS1302 实时时钟硬件仿真电路

10.4　程序设计

10.4.1　时钟读出程序设计

因为使用了时钟芯片 DS1302，时钟程序只需从 DS1302 各个寄存器中读出年、周、月、日、小时、分钟、秒等数据再处理即可，本次设计中仅读出时、分、秒数据。在首次对 DS1302 进行操作之前，必须对它进行初始化，然后从 DS1302 中读出数据，再经过处理后送给显示缓冲单元。时钟读出程序流程图见图 10.3。

10.4.2　时间调整程序设计

调整时间用 2 个调整按钮,1 个作为设定控制用,另外 1 个作为加调整用。在调整过程中,要调整的那位与别的位应该有区别,所以增加了闪烁功能,即调整的那位一直在闪烁直到调整下一位。闪烁原理就是让要调整的那一位,每隔一定时间熄灭一次,比如说50ms,利用定时器计时,当达到 50ms 时,就送给该位熄灭符,在下一次溢出时,再送正常显示的值,不断交替,直到调整该位结束。时间调整程序流程见图 10.4。

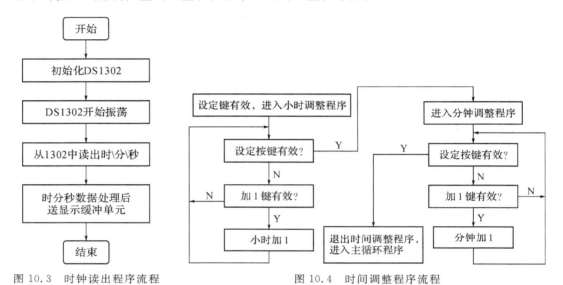

图 10.3　时钟读出程序流程　　　　　　　　　　图 10.4　时间调整程序流程

10.5　软件调试与运行结果

调试分为 Proteus 硬件电路调试和程序软件调试。硬件电路调试主要是检查各元件的连接线是否接好,另外可以通过编一个小调试软件来测试硬件电路是否正常。软件调试应分块进行,先进行显示程序调试,再写 DS1302 芯片的读写程序,最后通过多次修改与完善达到理想的功能效果。

DS1302 的晶振频率是计时精度的关键,实际设计中可换用标准晶振或用小电容进行修正,在本仿真电路中不需要对计时精度进行校准。

10.6　源程序清单

10.6.1　课程设计三参考汇编程序

```
;************************************************;
;              课程设计三程序:DS1302 实时时钟        ;
;                    12MHz 晶振                  ;
;************************************************;
;从 1302 中读出的数据放在 52H(小时)51H(分钟)50H(秒)
;显示缓冲单元：72H~73H(45H~44H)小时，
;75H-76H(43H-42H)分钟，
;78H-79H(41H-40H)秒
;定时器 T1 为调整时闪烁用
;显示式样：      15-38-12
;**************** 定义 ****************** ;
;
              SCLK        EQU P1.1      ;1302 时钟口,1302 第 7 脚
              IO          EQU P1.2      ;数据口,1302 第 6 脚
              RST         EQU P1.3      ;使能口,1302 第 5 脚
              KEYSW0      EQU P3.5      ;调时按键
              KEYSW1      EQU P3.6      ;加 1 按键
              BELL        EQU P3.7
              hour        DATA 52H      ;1302 读出时
              minute      DATA 51H      ;1302 读出分
              second      DATA 50H      ;1302 读出秒
              DS1302_ADDR DATA 32H      ;1302 需操作的地址数据存放
              DS1302_DATA DATA 31H      ;1302 读出或需写入的数据存放
              INTCON      DATA 30H      ;闪烁中断计时用
              CON_DATA    DATA 06H      ;闪烁时间 = 65ms×6 = 0.39s
              DISPFIRST   EQU 33H       ;显示地址首址
              DELAYR3     EQU 38H       ;延时程序用寄存器
              DELAYR5     EQU 39H       ;
              DELAYR6     EQU 3AH       ;
              DELAYR7     EQU 3BH       ;
;
;**************** 程序入口 ****************;
;
```

```
          ORG 0000H
          LJMP  START
          ORG 0003H
          RETI
          ORG 000BH
          RETI
          ORG 0013H
          RETI
          ORG 001BH
          LJMP  INTT1
          ORG 0023H
          RETI
          ORG 002BH
          RETI
;
;*****************;主程序;*********************;
;
          START:
                    MOV     SP,♯80H              ;堆栈在80H上
                    CLR     RST                  ;1302禁止
                    MOV     DISPFIRST,♯72H
                    MOV     74H,♯12              ;"-"
                    MOV     77H,♯12              ;"-"
                    MOV     TMOD,♯10H            ;计数器1,方式1
                    MOV     TL1,♯00H
                    MOV     TH1,♯00H
                    MOV     INTCON,♯CON_DATA
                    CLR     00H                  ;清闪烁标志
                    CLR     01H                  ;清闪烁标志
                    SETB    EA
                    MOV     DS1302_ADDR,♯8EH
                    MOV     DS1302_DATA,♯00H     ;允许写1302,♯80,禁止
                    LCALL   WRITE
                    MOV     DS1302_ADDR,♯90H
                    MOV     DS1302_DATA,♯0A6H
;1302充电电流1.1MA;♯A5：2.2MA;♯A7：0.6MA;
                    LCALL   WRITE
                    MOV     DS1302_ADDR,♯80H
                    MOV     DS1302_DATA,♯00H     ;1302晶振开始振荡,♯80H,禁止
```

```
                        LCALL   WRITE
;
;以下为主程序
            MAIN1：  MOV     DS1302_ADDR,♯85H  ;读出小时
                     LCALL   READ
                     MOV     hour,DS1302_DATA
                     LCALL   DISPLAY            ;显示刷新
                     MOV     DS1302_ADDR,♯83H  ;读出分钟
                     LCALL   READ
                     MOV     minute,DS1302_DATA
                     LCALL   DISPLAY            ;显示刷新
                     MOV     DS1302_ADDR,♯81H  ;读出秒
                     LCALL   READ
                     MOV     second,DS1302_DATA
                     LCALL   DISPLAY            ;显示刷新
                     MOV     R0,hour            ;小时分离,送显示缓存
                     LCALL   DIVIDE
                     MOV     73H,R1             ;时个位
                     MOV     44H,R1
                     MOV     72H,R2             ;时十位
                     MOV     45H,R2
                     LCALL   DISPLAY            ;显示刷新
                     MOV     R0,minute          ;分钟分离,送显示缓存
                     LCALL   DIVIDE
                     MOV     76H,R1             ;时个位
                     MOV     42H,R1
                     MOV     75H,R2             ;时十位
                     MOV     43H,R2
                     LCALL   DISPLAY            ;显示刷新
                     MOV     R0,second          ;秒分离,送显示缓存
                     LCALL   DIVIDE
                     MOV     79H,R1             ;秒个位
                     MOV     40H,R1
                     MOV     78H,R2             ;秒十位
                     MOV     41H,R2
                     LCALL   DISPLAY            ;显示刷新
;
                     JNB     KEYSW0,SETG        ;调整时间控制键
                     LJMP    MAIN1
```

```
;
;***************** 公历设置程序 *******************;
;
        SETG :
                RLCALL  DL20ms
                JB      KEYSW0,MAIN1
        WAITKEY0:  LCALL   DISPLAY              ;等待按键释放
                JNB     KEYSW0,WAITKEY0
                LCALL   DISPLAY
                JNB     KEYSW0,WAITKEY0
                MOV     78H,#00H              ;调时时秒位为 0
                MOV     79H,#00H              ;调时时秒位为 0
                MOV     40H,#00H              ;调时时秒位为 0
                MOV     41H,#00H              ;调时时秒位为 0
                MOV     DS1302_ADDR,#8EH
                MOV     DS1302_DATA,#00H      ;允许写 1302
                LCALL   WRITE
                MOV     DS1302_ADDR,#80H
                MOV     DS1302_DATA,#80H      ;1302 停止振荡
                LCALL   WRITE
                SETB    TR1                  ;闪烁开始
                SETB    ET1
        ;
        SETG9:     LCALL   DISPLAY              ;等待键按下
                JNB     KEYSW0,SETG10
                JNB     KEYSW1,GADDHOUR
                AJMP    SETG9
        GADDHOUR:  LCALL   DL20ms
                JB      KEYSW1,SETG9
                MOV     R7,52H               ;小时加 1
                LCALL   ADD1
                MOV     52H,A
                CJNE A,#24H,GADDHOUR11
        GADDHOUR11: JC     GADDHOUR1
                MOV     52H,#00H
        GADDHOUR1: MOV    DS1302_ADDR,#84H     ;小时值送入 1302
                MOV     DS1302_DATA,52H
                LCALL   WRITE
                MOV     R0,52H
```

```
                    LCALL    DIVIDE              ;小时值分离送显示缓存
                    MOV      73H,R1
                    MOV      44H,R1
                    MOV      72H,R2
                    MOV      45H,R2
        WAITKEY1:   LCALL    DISPLAY             ;等待按键释放
                    JNB      KEYSW1,WAITKEY1
                    LCALL    DISPLAY
                    JNB      KEYSW1,WAITKEY1
                    AJMP     SETG9
   ;

        SETG10:     LCALL    DL20ms
                    JB       KEYSW0,SETG9
                    SETB     01H                 ;调分时闪标志
        WAITKEY00:  LCALL    DISPLAY             ;等待按键释放
                    JNB      KEYSW0,WAITKEY00
                    LCALL    DISPLAY
                    JNB      KEYSW0,WAITKEY00
        SETG11:     LCALL    DISPLAY             ;等待分调整
                    JNB      KEYSW0,SETGOUT
                    JNB      KEYSW1,GADDMINTUE
                    AJMP     SETG11
   ;

        GADDMINTUE: LCALL    DL20ms
                    JB       KEYSW1,SETG11
                    MOV      R7,51H              ;分钟加 1
                    LCALL    ADD1
                    MOV      51H,A
                    CJNE     A,#60H,GADDMINTUE11
        GADDMINTUE11:JC      GADDMINTUE1
                    MOV      51H,#00H
        GADDMINTUE1:MOV      DS1302_ADDR,#82H    ;分钟值送入 1302
                    MOV      DS1302_DATA,51H
                    LCALL    WRITE
                    MOV      R0,51H
                    LCALL    DIVIDE              ;分钟值分离送显示缓存
                    MOV      76H,R1
                    MOV      42H,R1
                    MOV      75H,R2
```

```
                MOV      43H,R2
WAITKEY111：LCALL   DISPLAY              ;等待按键释放
                JNB      KEYSW1,WAITKEY111
                LCALL    DISPLAY
                JNB      KEYSW1,WAITKEY111
                AJMP     SETG11


SETGOUT：   LCALL    DL20ms
                JB       KEYSW0,SETG11
                MOV      DS1302_ADDR,#80H
                MOV      DS1302_DATA,#00H  ;1302 晶振开始振荡
                LCALL    WRITE
                MOV      DS1302_ADDR,#8EH
                MOV      DS1302_DATA,#80H  ;禁止写入 1302
                LCALL    WRITE
                CLR      00H
                CLR      01H
                CLR      ET1                  ;关闪中断
                CLR      TR1
WAITKEY000：LCALL   DISPLAY              ;等待按键释放
                JNB      KEYSW0,WAITKEY000
                LCALL    DISPLAY
                JNB      KEYSW0,WAITKEY000
                LJMP·    MAIN1
;
;*************** 闪动调时程序 *********************;
;
INTT1：     USH      ACC
                PUSH     PSW
                DJNZ     INTCON,GFLASHOUT
                MOV      INTCON,#CON_DATA
GFLASH：    CPL      00H
                JB       00H,GFLASH5
                MOV      72H,45H               ;全显示
                MOV      73H,44H
                MOV      75H,43H
                MOV      76H,42H
                MOV      78H,41H
                MOV      79H,40H
```

```
        GFLASHOUT:  LCALL   DISPLAY
                    POP     PSW
                    POP     ACC
                    RETI
;
        GFLASH5:    JB      01H,GFLASH6         ;调小时闪
                    MOV     72H,♯0AH
                    MOV     73H,♯0AH
                    AJMP    GFLASHOUT
        GFLASH6:    MOV     75H,♯0AH            ;调分钟闪
                    MOV     76H,♯0AH
                    AJMP    GFLASHOUT
;
;***************** 加 1 程序 ***************;
;
        ADD1:       MOV     A,R7
                    ADD     A,♯01H
                    DA      A
                    RET
;
;***************** 分离程序 *********************;
;
        DIVIDE:     MOV     A,R0
                    ANL     A,♯0FH
                    MOV     R1,A
                    MOV     A,R0
                    SWAP    A
                    ANL     A,♯0FH
                    MOV     R2,A
                    RET
;
;*************** 写 1302 程序 ***************;
;
        WRITE:      CLR     SCLK
                    NOP
                    SETB    RST
                    NOP
                    MOV     A,DS1302_ADDR
                    MOV     R4,♯8
```

```
        WRITE1：    RRC     A                        ;送地址给1302
                    NOP
                    NOP
                    CLR     SCLK
                    NOP
                    NOP
                    NOP
                    MOV     IO,C
                    NOP
                    NOP
                    NOP
                    SETB    SCLK
                    NOP
                    NOP
                    DJNZ    R4,WRITE1
                    CLR     SCLK
                    NOP
                    MOV     A,DS1302_DATA
                    MOV     R4,#8
        WRITE2：    RRC     A
                    NOP                              ;送数据给1302
                    CLR     SCLK
                    NOP
                    NOP
                    MOV     IO,C
                    NOP
                    NOP
                    NOP
                    SETB    SCLK
                    NOP
                    NOP
                    DJNZ    R4,WRITE2
                    CLR     RST
                    RET
;
;****************** 读1302程序 ******************;
;
        READ：      CLR     SCLK
                    NOP
```

```
              NOP
              SETB    RST
              NOP
              MOV     A,DS1302_ADDR
              MOV     R4,#8
READ1：       RRC     A                        ;送地址给 1302
              NOP
              MOV     IO,C
              NOP
              NOP
              NOP
              SETB    SCLK
              NOP
              NOP
              NOP
              CLR     SCLK
              NOP
              NOP
              DJNZ    R4,READ1

              MOV     R4,#8
READ2：       CLR     SCLK
              NOP                              ;从 1302 中读出数据
              NOP
              NOP
              MOV     C,IO
              NOP
              NOP
              NOP
              NOP
              NOP
              RRC     A
              NOP
              NOP
              NOP
              NOP
              SETB    SCLK
              NOP
              DJNZ    R4,READ2
```

```
                MOV     DS1302_DATA,A
                CLR     RST
                RET
;
;*************** 显示程序 *********************
        DISPLAY:    MOV     R1,DISPFIRST
                    MOV     R5,♯01H
        SPLAY:      MOV     A,R5
                    MOV     P2,A
                    MOV     A,@R1
                    MOV     DPTR,♯TABS
                    MOVC    A,@A+DPTR
                    MOV     P0,A
                    MOV     A,R5
                    LCALL   DL1ms
                    INC     R1
                    MOV     A,R5
                    JB      ACC.7,ENDOUTS
                    RL      A
                    MOV     R5,A
                    AJMP    SPLAY
        ENDOUTS:    MOV     P2,♯00H
                    MOV     P0,♯0FFH
                    RET
TABS: DB 0C0H,0F9H,0A4H,0B0H,99H,92H,82H,0F8H,80H,90H,0FFH,0C6H,0BFH,88H
;显示值  "0    1   2   3    4  5  6   7    8  9  不显 C   —    A"
;内存数  "0    1   2   3    4  5  6   7    8  9  0AH  0BH  0CH 0DH "
;*************** 延时子程序 *********************
;1ms 延时程序
        DL1ms:      MOV     DELAYR6,♯14H
        DL1:        MOV     DELAYR7,♯19H
        DL2:        DJNZ    DELAYR7,DL2
                    DJNZ    DELAYR6,DL1
                    RET
;20ms 延时程序
        DL20ms:     CLR     BELL
                    LCALL   DISPLAY
                    LCALL   DISPLAY
                    SETB    BELL
```

```
                         RET
;延时程序
        DL05S：      MOV      DELAYR3,♯20H
        DL05S1：     LCALL    DISPLAY
                     DJNZ     DELAYR3,DL05S1
                     RET


END
;********************* 结　束 *****************************
```

10.6.2　课程设计三参考 C 程序

```
/*-------------------------------------
Real－Time Clock DS1302 program V9.1
MCU STC89C52RC   XAL 12MHz
Build by Gavin Hu，2010.6.16
------------------------------------- */
♯ include ＜reg51.h＞
//
♯ define uchar unsigned char
♯ define uint unsigned int
♯ define ulong unsigned long
sbit BUZZ = P3^7;
sbit KEY1 = P3^5;
sbit KEY2 = P3^6;
sbit CE   = P1^3;
sbit SCLK = P1^1;
sbit IO   = P1^2;
uchar hour_reg, minute_reg, second_reg;


/*****************************************************/
/* Prototypes */
/*****************************************************/
uchar  rbyte_3w();
void   reset_3w();
void   wbyte_3w(uchar);
void read_time();
void delay(uint);
void display(uchar * );
```

```
void time2str(uchar * );
void time_set(void);

/*------------------------------------
   main function
------------------------------------ */
void main(void)
{
uchar dispram[9];
uchar s;
reset_3w();
wbyte_3w(0x8E);
wbyte_3w(0x00);
reset_3w();
wbyte_3w(0x90);
wbyte_3w(0xAB);
reset_3w();
wbyte_3w(0x81);
s = rbyte_3w();
reset_3w();
if (s&0x80)
    {
    wbyte_3w(0x80);
    wbyte_3w(s&0x7f);
    reset_3w();
    }
wbyte_3w(0x85);
s = rbyte_3w();
reset_3w();
if (s&0x80)
    {
    wbyte_3w(0x84);
    wbyte_3w(s&0x7f);
    reset_3w();
    }
while (1)
    {
    read_time();
    time2str(dispram);
```

```
        display(dispram);
        if (KEY1 == 0) time_set();
        }
    }


/*-------------------------------------
   Time data to display string function
   Parameter : pointer of string
   ------------------------------------- */
void time2str(uchar * ch)
{
ch[0] = hour_reg>>4;
ch[1] = hour_reg&0x0f;
ch[2] = 16;
ch[3] = minute_reg>>4;
ch[4] = minute_reg&0x0f;
ch[5] = 16;
ch[6] = second_reg>>4;
ch[7] = second_reg&0x0f;
}


/*-------------------------------------
   Set time function
   ------------------------------------- */
void time_set(void)
{
uchar ch[8];
uchar i,c;
reset_3w();
wbyte_3w(0x80);
wbyte_3w(0x80);
reset_3w();
second_reg = 0;
time2str(ch);
do {
    display(ch);
    } while (KEY1 == 0);
c = 2;
while (c)
```

```c
    {
    time2str(ch);
    if (c == 2) {ch[0]| = 0x80;ch[1]| = 0x80;}
        else {ch[3]| = 0x80;ch[4]| = 0x80;}
    display(ch);
    if (KEY1 == 0)
        {
        c -- ;
        do {
            display(ch);
            } while (KEY1 == 0);
        }
    if (KEY2 == 0)
        {
        if (c == 2)
            {
            hour_reg ++ ;
            if ((hour_reg&0x0f)>9) hour_reg = (hour_reg&0xf0) + 0x10;
            if (hour_reg>0x23) hour_reg = 0;
            }
            else
            {
            minute_reg ++ ;
            if ((minute_reg&0x0f)>9) minute_reg = (minute_reg&0xf0) + 0x10;
            if (minute_reg>0x59) minute_reg = 0;
            }
        for (i = 0;i<50;i ++ ) display(ch);
        }
    }
reset_3w();
wbyte_3w(0x84);
wbyte_3w(hour_reg);
reset_3w();
wbyte_3w(0x82);
wbyte_3w(minute_reg);
reset_3w();
wbyte_3w(0x80);
wbyte_3w(0x00);
reset_3w();
}
```

```c
/*---------------------------------------
   Delay function
   Parameter: unsigned int dt
   Delay time = dt(ms)
   --------------------------------------- */
void delay(unsigned int dt)
{
register unsigned char bt,ct;
for (;dt;dt --)
    for (ct = 2;ct;ct --)
        for (bt = 248; -- bt;);
}


/*---------------------------------------
   8 LED digital tubes display function
   Parameter: sting pointer to display
   --------------------------------------- */
void display(uchar * disp_ram)
{
static uchar disp_count;
unsigned char i,j;
unsigned char code table[] =
{0xc0,0xf9,0xa4,0xb0,0x99,0x92,0x82,0xf8,0x80,0x90,0x88,
0x83,0xc6,0xa1,0x86,0x8e,0xbf,0xff};
disp_count = (disp_count + 1)&0x7f;
for (i = 0;i<8;i ++ )
    {
    j = disp_ram[i];
    if (j&0x80) P0 = (disp_count>32)? table[j&0x7f]:0xff;
        else P0 = table[j];
    P2 = 0x01<<i;
    delay(1);
    P0 = 0xff;
    P2 = 0;
    }
}
```

```
/*---------------------------------------
   Read time function
------------------------------------ */
void read_time()
{
reset_3w();
wbyte_3w(0xBF);
second_reg = rbyte_3w()&0x7f;
minute_reg = rbyte_3w()&0x7f;
hour_reg = rbyte_3w()&0x3f;
reset_3w();
}


void reset_3w() /*----- reset and enable the 3 - wire interface ------ */
{
    CE = 0;
    SCLK = 0;
    CE = 0;
    SCLK = 0;
    CE = 1;
}


void wbyte_3w(uchar W_Byte) /*------ write one byte to the device ------- */
{
uchar i;
    for(i = 0;i < 8; ++ i)
    {
        SCLK = 0;
        IO = W_Byte & 0x01;
        SCLK = 1;
        W_Byte >> = 1;
    }
}
uchar rbyte_3w() /*------- read one byte from the device -------- */
{
uchar i;
uchar R_Byte;
    IO = 1;
    for(i = 0;i < 8;i ++ )
```

```
    {
        SCLK = 0;
        R_Byte >> = 1;
        if (IO)   R_Byte | = 0x80;
        SCLK = 1;
    }
    return R_Byte;
}
```

第 11 章

课程设计四:数字温度计

11.1 系统功能

数字温度计测温范围在$-55\sim125$℃,精度误差在 0.5℃以内,用四位共阳 LED 数码管直读显示,要求高位为 0℃时不显示,低于 0℃时前面显示"$-$"。

11.2 设计方案

传统的测温元件有热电偶和热电阻,而热电偶和热电阻测出的一般都是电压,再转换成对应的温度,需要比较多的外部硬件支持,硬件电路复杂,软件调试复杂,制作成本高。而数字温度计的设计可采用美国 DALLAS 半导体公司继 DS1820 之后推出的一种改进型智能温度传感器 DS18B20 作为检测元件,测温范围为$-55\sim125$℃,分辨率最大可达 0.0625℃。DS18B20 可以直接读出被测温度值(不用校准),而且采用单线与单片机通讯,减少了外部的硬件电路,具有高精度和易使用的特点。

按照系统功能的要求,数字温度计由主控制器、测温单元、显示电路共 3 个模块组成。总体系统结构详见图 11.1。

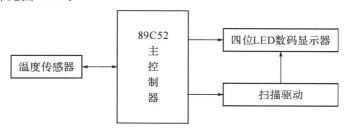

图 11.1 数字温度计系统结构

11.3　系统硬件仿真电路

数字温度计硬件仿真电路见图 11.2，控制器使用 89C52 系列单片机，温度传感器使用 DS18B20，用四位共阳 LED 数码管以动态扫描法实现温度显示，从 P0 口输出段码，列扫描用 P2 口来实现，列驱动用 74HC244，可直接作为 LED 段码灯的电源。

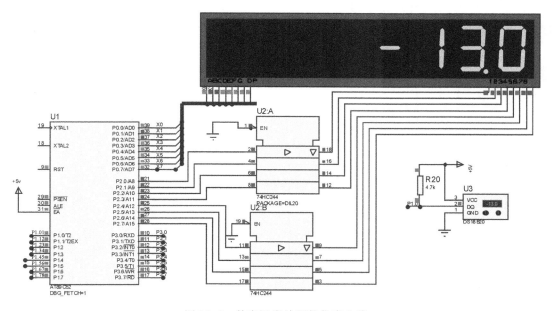

图 11.2　数字温度计硬件仿真电路

11.4　程序设计

系统程序主要包括主程序、读出温度子程序、温度转换命令子程序、计算温度子程序、显示数据刷新子程序等。

11.4.1　主程序

主程序的主要功能是负责温度的实时显示、读出并处理 DS18B20 的测量温度值。温度测量每 1s 进行一次，其程序流程详见图 11.3。

11.4.2　读出温度子程序

读出温度子程序的主要功能是读出 DS18B20 RAM 中的 9 个字节,在读出时需进行 CRC 校验,校验有错时不进行温度数据的改写,其程序流程详见图 11.4。

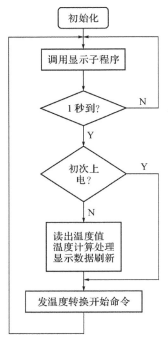

图 11.3　DS18B20 温度计主程序流程

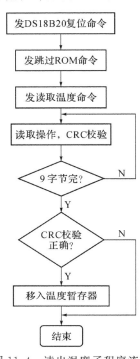

图 11.4　读出温度子程序流程

11.4.3　温度转换命令子程序

温度转换命令子程序主要是发温度转换开始命令,当采用 12 位分辨率时,转换时间约为 750ms,在本程序设计中采用 1s 显示程序延时法等待转换的完成。温度转换命令子程序流程详见图 11.5。

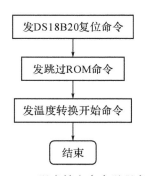

图 11.5　温度转变命令子程序流程

11.4.4　计算温度子程序

计算温度子程序将 DS18B20 RAM 中读取值进行 BCD 码的转换运算,并进行温度值正负的判定,其程序流程详见图 11.6。

11.4.5 显示数据刷新子程序

显示数据刷新子程序主要是对显示缓冲器中的显示数据进行刷新操作,当最高数据显示位为零时将符号显示位移入下一位,程序流程图见图 11.7。

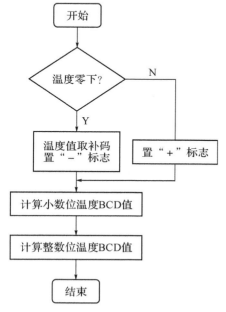

图 11.6 计算温度子程序流程

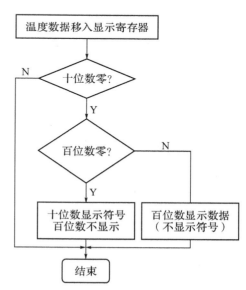

图 11.7 显示数据刷新子程序流程

11.4.6 DS18B20 中的 ROM 命令

1. Read ROM [33H]

这个命令允许总线控制器读到 DS18B20 的 8 位系列编码、唯一的序列号和 8 位 CRC 码。只有在总线上存在单只 DS18B20 时才能使用这个命令。如果总线上有不止一个从机,当所有从机试图同时传送信号时就会发生数据冲突(漏极开路连在一起形成相与的效果)。

2. Match ROM [55H]

这是匹配 ROM 命令,后跟 64 位 ROM 序列,让总线控制器在多点总线上定位一只特定的 DS18B20。只有和 64 位 ROM 序列完全匹配的 DS18B20 才能响应随后的存储器操作。所有和 64 位 ROM 序列不匹配的从机都将等待复位脉冲。这条命令在总线上有单个或多个器件时都可以使用。

3. Skip ROM［0CCH］

这条命令允许总线控制器不用提供 64 位 ROM 编码就可以使用存储器操作命令，在单点总线情况下，可以节省时间。如果总线上不止一个从机，在 Skip ROM 命令之后跟着发一条读命令，由于多个从机同时传送信号。总线上就会发生数据冲突（漏极开路下拉效果相当于相与）。

4. Search ROM［0F0H］

当一个系统初次启动时，总线控制器可能并不知道单线总线上有多少器件或它们的 64 位 ROM 编码。搜索 ROM 命令允许总线控制器用排除法识别总线上的所有从机的 64 位编码。

5. Alarm Search［0ECH］

这条命令的流程和 Search ROM 相同。然而，只有在最近一次测温后遇到符合报警条件的情况，DS18B20 才会响应这条命令。报警条件定义为温度高于 TH 或低于 TL。只要 DS18B20 不掉电，报警状态将一直保持，直到再一次测得的温度值达不到报警条件。

6. Write Scratchpad［4EH］

这个命令向 DS18B20 的暂存器 TH 和 TL 中写入数据。可以在任何时刻发出复位命令来中止写入。

7. Read Scratchpad［0BEH］

这个命令读取暂存器的内容。读取将从第 1 个字节开始，一直进行下去，直到第 9（CRC）个字节读完。如果不想读完所有字节，控制器可以在任何时间发出复位命令来中止读取。

8. Copy Scratchpad［48H］

这个命令把暂存器的内容拷贝到 DS18B20 的 E²PROM 存储器里，即把温度报警触发字节存入非易失性存储器里。如果总线控制器在这条命令之后跟着发出读时间隙，而 DS18B20 又忙于把暂存器拷贝到 E²PROM 存储器，DS18B20 就会输出一个"0"，如果拷贝结束的话，DS18B20 则输出"1"。如果使用寄生电源，总线控制器必须在这条命令发出后立即启动强上拉并最少保持 10ms。

9. Convert T［44H］

这条命令启动一次温度转换而无需其他数据。温度转换命令被执行，而后 DS18B20 保持等待状态。如果总线控制器在这条命令之后跟着发出读时间隙，而 DS18B20 又忙于做时间转换的话，DS18B20 将在总线上输出"0"，若温度转换完成，则输出"1"。如果使用寄生电源，总线控制器必须在发出这条命令后立即启动强上拉，并保持 500ms 以上时间。

10. Recall E²[0B8H]

这条命令把报警触发器里的值拷贝回暂存器。这种拷贝操作在 DS18B20 上电时自动执行,这样器件一上电暂存器里马上就存在有效的数据了。若在这条命令发出之后发出读数据隙,器件会输出温度转换忙的标识:"0"=忙,"1"=完成。

11. Read Power Supply [0B4H]

若把这条命令发给 DS18B20 后发出读时间隙,器件会返回它的电源模式:"0"=寄生电源,"1"=外部电源。

11.4.7　温度数据的计算处理方法

从 DS18B20 读取出的二进制值必须先转换成十进制 BCD 码,才能用于字符的显示。因为 DS18B20 的转换精度为 9~12 位(可选的),为了提高精度可采用 12 位。在采用 12 位转换精度时,温度寄存器里的值是以 0.0625 为步进的,即温度值为温度寄存器里的二进制值乘以 0.0625,就是实际的十进制温度值。表 11.1 是 DS18B20 温度与二进制及十六进制表示值的对应关系,从表中可知,一个十进制温度值和二进制值之间有很明显的关系,就是把二进制的高字节的低半字节和低字节的高半字节组成一个字节,这个字节的二进制值转化为十进制 BCD 码值后,就是温度值的百、十、个位值,而剩下的低字节的低半字节转化成十进制后,就是温度值的小数部分。小数部分因为是半个字节,所以十六进制值范围是 0~F,转换成十进制小数值就是 0.0625 的倍数(0~15 倍)。需用 4 位的数码管来显示小数部分,在实际应用中不必有这么高的精度,设计中一般采用 1 位数码管来显示小数,可以精确到 0.1℃。表 11.2 是小数部分十六进制和十进制的近似对应关系表。

表 11.1　DS18B20 温度与表示值对应关系

温度(℃)	二进制表示		十六进制表示
+125	0000 0111	1101 0000	07D0H
+85	0000 0101	0101 0000	0550H
+25.0625	0000 0001	1001 0001	0191H
+10.125	0000 0000	1010 0010	00A2H
+0.5	0000 0000	0000 1000	0008H
0	0000 0000	0000 0000	0000H
−0.5	1111 1111	1111 1000	FFF8H
−10.125	1111 1111	0101 1110	FF5EH
−25.0625	1111 1110	0110 1111	FE6FH
−55	1111 1100	1001 0000	FC90H

表 11.2　小数部分十六进制和十进制的近似对应关系

小数部分 十六进制值	0	1	2	3	4	5	6	7	8	9	A	B	C	D	E	F
十进制 小数近似值	0	0	1	1	2	3	3	4	5	5	6	6	7	8	8	9

11.5　软件调试与运行结果

　　系统的调试以程序为主。可先编一个测试小程序以判断仿真硬件电路是否正常。然后分别进行显示程序、主程序、读出温度子程序、温度转换命令子程序、计算温度子程序、显示数据刷新等子程序的编程及调试,由于 DS18B20 与单片机采用单线数据传送,因此,对 DS18B20 进行读写编程时必须严格地保证读写时序,否则将无法读取测量结果。

11.6　源程序清单

11.6.1　课程设计四参考汇编程序

```
;*************************************************************
;                课程设计四程序:数字温度计                   *
;            显示精度为 0.1℃ ,测温范围 - 55～ + 125℃         *
;            用 89C52 系列单片机,12MHz 晶振                   *
;*************************************************************
;
;*************************************************************
;
;        常数定义
;
;*************************************************************
     TIMEL      EQU     0E0H        ;定时器 T0 的 20ms 时间常数
     TIMEH      EQU     0B1H        ;定时器 T0 的 20ms 时间常数
     TEMPHEAD   EQU     36H         ;18B20 读出字节存放首址(共读 9 个字节)
;
;*************************************************************
;
;        工作内存定义
;
;*************************************************************
```

```
        BITST       DATA     20H          ;用作标志位
        TIME1SOK    BIT      BITST.1      ;1s 定时时间标志,1s 到时为 1
        TEMPONEOK   BIT      BITST.2      ;上电标志,刚上电为 0,读出一次后为 1
        TEMPL       DATA     26H          ;读出温度低字节存放——整数低四位 + 小数位四位
        TEMPH       DATA     27H          ;读出温度高字节存放——四位符号位 + 整数高四位
        TEMPHC      DATA     28H          ;用于存放处理好的 BCD 码温度值:百位 + 十位
        TEMPLC      DATA     29H          ;用于存放处理好的 BCD 码温度值:个位 + 小数位
    ;
    ;
    ;*************************************************************
    ;   引脚定义
    ;*************************************************************
        TEMPDIN     BIT      P1.0         ;18B20 数据接口
    ;*************************************************************
    ;   中断向量区
    ;*************************************************************
        ORG         0000H
        LJMP        START
        ORG         00BH
        LJMP        T0IT
    ;*************************************************************
    ;   系统初始化
    ;*************************************************************
        ORG         100H
START:      MOV     SP, #60H
CLSMEM:     MOV     R0, #20H            ;堆栈底
            MOV     R1, #60H            ;20H~7FH 清零
CLSMEM1:    MOV     @R0, #00H           ;
            INC     R0                  ;
            DJNZ    R1, CLSMEM1         ;
    ;
            MOV     TMOD, #00100001B    ;定时器 0 作方式 1 (16BIT)
            MOV     TH0, #TIMEL         ;装 20ms 定时初值
            MOV     TL0, #TIMEH         ;装 20ms 定时初值
            MOV     P2, #00H            ;LED 显示关
            SJMP    INIT
    ;
ERROR:      NOP
            LJMP    START
```

```
;
                NOP
INIT:           NOP
                SETB        ET0
                SETB        TR0
                SETB        EA
                MOV         PSW，#00H
                CLR         TEMPONEOK              ;第一次上电为 0
                LJMP        MAIN                   ;
;
;*************************************************************;
;       定时器 0 中断服务程序                                  ;
;*************************************************************;
T0IT:           PUSH        PSW
                MOV         PSW，#10H
                MOV         TH0，#TIMEH
                MOV         TL0，#TIMEL
                INC         R7
                CJNE        R7，#32H，T0IT1
                MOV         R7，#00H
                SETB        TIME1SOK               ;1s 定时到标志位为 1
T0IT1:          POP         PSW
                RETI
;
;*************************************************************;
;       主程序                                                ;
;*************************************************************;

MAIN:           LCALL       DISP1                  ;调用显示子程序
                JNB         TIME1SOK，MAIN         ;
                CLR         TIME1SOK               ;测温每 1s 一次
                JNB         TEMPONEOK，MAIN2       ;上电时先温度转换一次
                LCALL       READTEMP1              ;读出温度值子程序
                LCALL       CONVTEMP               ;温度 BCD 码计算处理子程序
                LCALL       DISPBCD                ;显示区 BCD 码温度值刷新子程序
                LCALL       DISP1                  ;消闪烁,显示一次
MAIN2:          LCALL       READTEMP               ;温度转换开始
                SETB        TEMPONEOK              ;
                LJMP        MAIN                   ;
```

```
;
;***********************************************************;
;        以下为子程序区                                      ;
;***********************************************************;
;        DS18B20 复位子程序                                  ;
;***********************************************************;
INITDS1820:    SETB     TEMPDIN
               NOP
               NOP
               CLR      TEMPDIN
               MOV      R6，#0A0H          ;DELAY 480us
               DJNZ     R6，$
               MOV      R6，#0A0H
               DJNZ     R6，$
               SETB     TEMPDIN
               MOV      R6，#32H           ;DELAY 70us
               DJNZ     R6，$
               MOV      R6，#3CH
LOOP1820:      MOV      C，TEMPDIN
               JC       INITDS1820OUT
               DJNZ     R6，LOOP1820
               MOV      R6，#064H          ;DELAY 200us
               DJNZ     R6，$
               SJMP     INITDS1820
               RET
;
INITDS1820OUT: SETB     TEMPDIN
               RET
;
;***********************************************************;
;     读 DS18B20 的程序，在 DS18B20 中读出一个字节的数据       ;
;***********************************************************;
READDS1820:    MOV      R7，#08H
               SETB     TEMPDIN
               NOP
               NOP
READDS1820LOOP:CLR      TEMPDIN
               NOP
               NOP
```

```
                NOP
                SETB      TEMPDIN
                MOV       R6, ♯07H              ;DELAY 15us
                DJNZ      R6, $
                MOV       C, TEMPDIN
                MOV       R6, ♯3CH              ;DELAY 120us
                DJNZ      R6, $
                RRC       A
                SETB      TEMPDIN
                DJNZ      R7, READDS1820LOOP
                MOV       R6, ♯3CH              ;DELAY 120us
                DJNZ      R6, $
                RET
;
;
;*********************************************************;
;      写 DS18B20 的程序,在 DS18B20 中写一个字节的数据        ;
;*********************************************************;
WRITEDS1820:    MOV       R7, ♯08H
                SETB      TEMPDIN
                NOP
                NOP
WRITEDS1820LOP:CLR        TEMPDIN
                MOV       R6, ♯07H              ;DELAY 15us
                DJNZ      R6, $
                RRC       A
                MOV       TEMPDIN, C
                MOV       R6, ♯34H              ;DELAY 104us
                DJNZ      R6, $
                SETB      TEMPDIN
                DJNZ      R7, WRITEDS1820LOP
                RET
;
;*********************************************************;
;      以下为读温度子程序                                    ;
;*********************************************************;
READTEMP:       LCALL     INITDS1820            ;温度转换开始命令程序
                MOV       A, ♯0CCH
                LCALL     WRITEDS1820           ;SKIP ROM
```

```
                MOV       R6，♯34H            ;DELAY 104us
                DJNZ      R6，$
                MOV       A，♯44H
                LCALL     WRITEDS1820        ;START CONVERSION
                MOV       R6，♯34H            ;DELAY 104us
                DJNZ      R6，$
                RET
;
READTEMP1：      LCALL     INITDS1820         ;读出温度字节子程序
                MOV       A，♯0CCH
                LCALL     WRITEDS1820        ;SKIP ROM
                MOV       R6，♯34H            ;DELAY 104us
                DJNZ      R6，$
                MOV       A，♯0BEH
                LCALL     WRITEDS1820        ;SCRATCHPAD
                MOV       R6，♯34H            ;DELAY 104us
                DJNZ      R6，$
                MOV       R5，♯09H            ;读 9 个字节
                MOV       R0，♯TEMPHEAD
                MOV       B，♯00H
READTEMP2：      LCALL     READDS1820
                MOV       @R0，A
                INC       R0
READTEMP21：     LCALL     CRC8CAL            ;CRC 校验
                DJNZ      R5，READTEMP2
                MOV       A，B
                JNZ       READTEMPOUT        ;校验出错结束
                MOV       A，TEMPHEAD + 0     ;校验正确,将前 2 个字节(温度)暂存
                MOV       TEMPL，A
                MOV       A，TEMPHEAD + 1
                MOV       TEMPH，A
READTEMPOUT：    RET
;
;
;***********************************************************;
;      处理温度 BCD 码子程序                                 ;
;***********************************************************;
CONVTEMP：       MOV       A，TEMPH
                ANL       A，♯80H
```

```
            JZ        TEMPC1
            CLR       C                              ;负温度的补码处理(＋1后取反)
            MOV       A, TEMPL
            CPL       A
            ADD       A, ＃01H
            MOV       TEMPL, A
            MOV       A, TEMPH
            CPL       A
            ADDC      A, ＃00H
            MOV       TEMPH, A              ;TEMPHC HI＝符号位
            MOV       TEMPHC, ＃0BH
            SJMP      TEMPC11
;
TEMPC1：    MOV       TEMPHC, ＃0AH         ;正温度处理
TEMPC11：   MOV       A, TEMPHC
            SWAP      A
            MOV       TEMPHC, A
            MOV       A, TEMPL              ;
            ANL       A, ＃0FH              ;温度小数位处理(乘0.0625)
            MOV       DPTR, ＃TEMPDOTTAB    ;用查表处理小数位
            MOVC      A, @A＋DPTR           ;
            MOV       TEMPLC, A             ;TEMPLC LOW＝小数部分 BCD
;
            MOV       A, TEMPL             ;处理温度整数部分
            ANL       A, ＃0F0H             ;
            SWAP      A                     ;
            MOV       TEMPL, A              ;
            MOV       A, TEMPH              ;
            ANL       A, ＃0FH              ;
            SWAP      A                     ;
            ORL       A, TEMPL              ;
            LCALL     HEX2BCD1              ;
            MOV       TEMPL, A              ;
            ANL       A, ＃0F0H             ;
            SWAP      A                     ;
            ORL       A, TEMPHC             ;TEMPHC LOW 放十位数 BCD
            MOV       TEMPHC, A             ;
            MOV       A, TEMPL              ;
            ANL       A, ＃0FH              ;
```

```
              SWAP      A                      ;TEMPLC HI 放个位数 BCD
              ORL       A, TEMPLC              ;
              MOV       TEMPLC, A              ;
              MOV       A, R7                  ;
              JZ        TEMPC12                ;
              ANL       A, ♯0FH                ;
              SWAP      A                      ;
              MOV       R7, A                  ;
              MOV       A, TEMPHC              ;TEMPHC HI 放百位数 BCD
              ANL       A, ♯0FH                ;
              ORL       A, R7                  ;
              MOV       TEMPHC, A              ;
TEMPC12:      RET                             ;
;
;************************************************************;
;         小数部分码表                                      ;
;************************************************************;
TEMPDOTTAB:   DB  00H, 01H, 01H, 02H, 03H, 03H, 04H, 04H, 05H, 06H
;
              DB  06H, 07H, 08H, 08H, 09H, 09H
;
              RET
;
;************************************************************;
;         显示区 BCD 码温度值刷新子程序                       ;
;************************************************************;
;低于零摄氏度要显示"-",高位零不显示
DISPBCD:      MOV       A, TEMPLC              ;
              ANL       A, ♯0FH               ;
              MOV       70H, A                 ;
              MOV       A, TEMPLC              ;
              SWAP      A                      ;
              ANL       A, ♯0FH               ;
              MOV       71H, A                 ;
              MOV       A, TEMPHC              ;
              ANL       A, ♯0FH               ;
              MOV       72H, A                 ;
              MOV       A, TEMPHC              ;
              SWAP      A                      ;
```

```
              ANL       A，♯0FH               ;
              MOV       73H，A                ;
              MOV       A，TEMPHC             ;
              ANL       A，♯0F0H              ;
              CJNE      A，♯010H，DISPBCD0 ;
              SJMP      DISPBCD2             ;
;
DISPBCD0：     MOV       A，TEMPHC             ;
              ANL       A，♯0FH               ;
              JNZ       DISPBCD2                      ;十位数是零
              MOV       A，TEMPHC             ;
              SWAP      A                    ;
              ANL       A，♯0FH               ;
              MOV       73H，♯0AH                      ;符号位不显示
              MOV       72H，A                         ;十位数显示符号
DISPBCD2：     RET       ;
;
;**********************************************************;
;                    显示子程序                             ;
;**********************************************************;
;显示数据在 70H～73H 单元内,用 4 位共阳数码管,P0 口输出段码数据,
;P2 口作扫描控制,每个 LED 数码管亮 1ms 时间再逐位循环
;
DISP1：        MOV       R1，♯70H                       ;指向显示数据首址
              MOV       R5，♯80H                       ;扫描控制字初值
PLAY：         MOV       P0，♯0FFH             ;
              MOV       A，R5                          ;扫描字放入 A
              MOV       P2，A                          ;从 P2 口输出
              MOV       A，@R1                         ;取显示数据到 A
              MOV       DPTR，♯TAB                     ;取段码表地址
              MOVC      A，@A+DPTR                     ;查显示数据对应段码
              MOV       P0，A                          ;段码放入 P0 口
              MOV       A，R5                 ;
              JNB       ACC.6，LOOP5                   ;小数点处理
              CLR       P0.7                 ;
LOOP5：        LCALL     DL1ms                         ;显示 1ms
              INC       R1                            ;指向下一地址
              MOV       A，R5                          ;扫描控制字放入 A
              JB        ACC.4，ENDOUT                  ;ACC.5＝0 时一次显示结束
```

```
              RR        A                      ;A 中数据循环左移
              MOV       R5,A                   ;放回 R5 内
              AJMP      PLAY                   ;跳回 PLAY 循环
ENDOUT：      MOV       P0,#0FFH               ;一次显示结束,P0 口复位
              MOV       P2,#00H                ;P2 口复位
              RET                              ;子程序返回
TAB：    DB  0C0H,0F9H,0A4H,0B0H,99H,92H,82H,0F8H,80H,90H,0FFH,0BFH
;共阳段码表 "0"" 1"" 2" "3" "4"" 5" "6"" 7" "8" "9""不亮" "-"
;
;1ms 延时程序(LED 显示程序用)
      DL1ms：  MOV       R6,#14H
      DL1：    MOV       R7,#19H
      DL2：    DJNZ      R7,DL2
              DJNZ      R6,DL1
              RET
;
;***********************************************************;
;        单字节十六进制转 BCD                              ;
;***********************************************************;
HEX2BCD1：    MOV       B,#064H                ;十六进制 -> BCD
              DIV       AB                     ;B = A % 100
              MOV       R7,A                   ;R7 = 百位数
              MOV       A,#0AH
              XCH       A,B
              DIV       AB                     ;B = A % B
              SWAP      A
              ORL       A,B
              RET
;
;
;***************************************************************;
;        CRC 校验程序                                         ;
;        X^8 + X^5 + X^4 + 1                                  ;
;***************************************************************;

CRC8CAL：     PUSH      ACC
              MOV       R7,#08H
;
CRC8LOOP1：   XRL       A,B
```

```
                RRC     A
                MOV     A, B
                JNC     CRC8LOOP2
                XRL     A, #18H
;
CRC8LOOP2:      RRC     A
                MOV     B, A
                POP     ACC
                RR      A
                PUSH    ACC
                DJNZ    R7, CRC8LOOP1
                POP     ACC
                RET
;
                END                     ;程序结束
```

11.6.2 课程设计四参考 C 程序

```c
/*-------------------------------------
DS18B20 Digital Thermometer program V10.1
MCU STC89C52RC   XAL 12MHz
Build by Gavin Hu, 2010.6.14
------------------------------------- */
# include "reg51.h"
# include "intrins.h"
# define NO_DISPLAY 17
# define DISP_SIGN 16
sbit ONE_WIRE_DQ = P1^0;

void delay_ms(unsigned int);
void delay_us(register unsigned char);
void temp2str(signed int tmep, unsigned char * );
void display(unsigned char * );
void start_convert(void);
signed int read_temperature(void);
unsigned char OW_reset(void);
unsigned char OW_read_byte(void);
void OW_write_byte(unsigned char val);
void main()
{
```

```c
unsigned char i;
unsigned char dispram[8];
for (i = 0; i<8; i ++ ) dispram[i] = NO_DISPLAY;
while (1)
    {
    start_convert();
    for (i = 0; i<120; i ++ ) display(dispram);
    temp2str(read_temperature(),dispram);
    }
}
/*-----------------------------------
// Start DS18B20 Temperature Convert
----------------------------------- */
void start_convert(void)
{
OW_reset();
OW_write_byte(0xCC);//Skip ROM
OW_write_byte(0x44);// Start Conversion
}
/*-----------------------------------
// Read Temperature
// returns the Temperature.
----------------------------------- */
signed int read_temperature(void)
{
unsigned char get[9];
signed int temp;
unsigned char i;
OW_reset();
OW_write_byte(0xCC);                                // Skip ROM
OW_write_byte(0xBE);                                // Read Scratch Pad
for (i = 0; i<9; i ++ ) get[i] = OW_read_byte();
temp = get[1];                                      // Sign byte + 1sbit
temp = (temp<<8) | get[0];                          // Temp data plus lsb
return temp;
}
/*-----------------------------------
// OW_RESET-performs a reset on the one-wire bus and
// returns the presence detect.
----------------------------------- */
```

```c
unsigned char OW_reset(void)
{
unsigned char presence;
ONE_WIRE_DQ = 0;                   //pull ONE_WIRE_DQ line low
delay_us(240);                     // leave it low for 480us
ONE_WIRE_DQ = 1;                   // allow line to return high
delay_us(33);                      // wait for presence 70us
presence = ! ONE_WIRE_DQ;          // get presence signal
delay_us(205);                     // wait for end of timeslot
return (presence);                 // presence signal returned
}                                  // 0 = presence, 1 = no part


/*----------------------------------------
// READ_BYTE-reads a byte from the one-wire bus.
---------------------------------------- */
unsigned char OW_read_byte(void)
{
unsigned char i;
unsigned char value;
for (i = 0;i<8;i++ )
    {
    ONE_WIRE_DQ = 0;               // pull ONE_WIRE_DQ low to

start timeslot
    value >> = 1;                  // delay
    ONE_WIRE_DQ = 1;               // then return high
    delay_us(1);                   // delay 6us from start of

timeslot
    if (ONE_WIRE_DQ) value|= 0x80; // reads byte in, one byte at a time and then
    delay_us(25);                  // delay 55us
    }                              //delay_us(60);
return(value);
}
/*----------------------------------------
// WRITE_BYTE-writes a byte to the one-wire bus.
---------------------------------------- */
void OW_write_byte(char val)
{
unsigned char i;
```

```
    for ( i = 0 ; i<8 ; i ++ )                    // writes byte, one bit at a time
        {
        ONE_WIRE_DQ = 0 ;                         // pull ONE_WIRE_DQ low to
                                                  //start timeslot

        delay_us(1) ;                             // delay 6us
        ONE_WIRE_DQ = val&0x01 ;                  // return ONE_WIRE_DQ high

        if write 1
            delay_us(30) ;                        // hold value for remainder of timeslot 64us
        ONE_WIRE_DQ = 1 ;
        val >> = 1 ;                              // shifts val right 1 spaces
        }
    }
/*----------------------------------------
   Temperature data to display string function
   Parameter : int temp, pointer of string
---------------------------------------- */
void temp2str(signed int temp, unsigned char * ch)
{
unsigned char sign ;
if ( temp<0 )
    {
    sign = 1 ;
    temp = ( ~temp) + 1 ;
    }
    else sign = 0 ;
ch[7] = ((temp&0x000f) * 10 + 8)/16 ;
temp >> = 4 ;
ch[6] = (temp % 10)|0x40 ;
temp / = 10 ;
ch[5] = temp % 10 ;
temp / = 10 ;
ch[4] = temp % 10 ;
ch[3] = NO_DISPLAY ;
if ( ch[4] == 0)
    {
    ch[4] = NO_DISPLAY ;
    if ( ch[5] == 0) ch[5] = NO_DISPLAY ;
    }
```

```c
if (sign)
    {
    if (ch[5] == NO_DISPLAY) ch[5] = DISP_SIGN;
        else if (ch[4] == NO_DISPLAY) ch[4] = DISP_SIGN;
        else ch[3] = DISP_SIGN;
    }
}
/*------------------------------------
  8 LED digital tubes display function
  Parameter: sting pointer to display
  ------------------------------------ */
void display(unsigned char * disp_ram)
{
static unsigned char disp_count;
unsigned char i;
unsigned char code table[] =
{0xc0,0xf9,0xa4,0xb0,0x99,0x92,0x82,0xf8,0x80,0x90,0x88,
0x83,0xc6,0xa1,0x86,0x8e,0xbf,0xff};
disp_count = (disp_count + 1)&0x7f;
for (i = 0;i<8;i ++ )
    {
    if (disp_ram[i]&0x80) P0 = (disp_count>32)? table [disp_ram[i]&0x3f]:0xff;
        else P0 = table[disp_ram[i]&0x3f];
    if (disp_ram[i]&0x40) P0 & = 0x7f;
    P2 = 0x01<<i;
    delay_ms(1);
    P0 = 0xff;
    P2 = 0;
    }
}


/*------------------------------------
  Delay function
  Parameter: unsigned char dt
  Delay time = dt * 2 + 5(us)
  ------------------------------------ */
void delay_us(register unsigned char dt)
{
while (-- dt);
```

```
}


/*-------------------------------------
  Delay function
  Parameter: unsigned int dt
  Delay time = dt(ms)
-------------------------------- */
void delay_ms(unsigned int dt)
{
register unsigned char bt,ct;
for (;dt;dt --)
    for (ct = 2;ct;ct --)
        for (bt = 250; -- bt;);
}
```

第 12 章

课程设计五：低频信号发生器

12.1 系统功能

低频信号发生器要求能输出 0.1～50Hz 的正弦波、三角波信号，其中正弦波和三角波信号可以用按键选择输出，输出信号的频率可以在 0.1～50Hz 的范围内调整。

12.2 设计方案

因输出信号的频率较低，可使用单片机作为信号数据产生源，用中断查表法完成波形数据的输出，再用 DA 转换器输出规定的波形信号。另外也可利用多余的端口经 DA 转换输出 0°～360°的移相波形，同时也可输出一路方波信号。系统实现的结构框图见图 12.1。

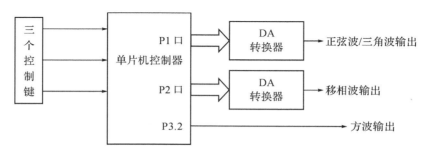

图 12.1　低频信号源系统结构

12.3　系统硬件仿真电路

低频信号源硬件仿真电路见图 12.2。

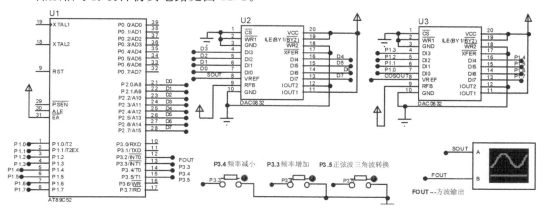

图 12.2　低频信号源硬件仿真电路

12.3.1　控制部分

控制芯片选择 89C52 系列单片机。P3.3～P3.5 口接三个按键，其中 P3.3 口按键为频率增加键，P3.4 口按键为频率减小键，P3.5 口按键为正弦波与三角波选择按键。P1口输出正弦波或三角波数据，P2 口输出移相波数据，P3.2 输出方波。

12.3.2　数模(D/A)转换部分

DAC0832 是 CMOS 工艺制造的 8 位 D/A 转换器，属于 8 位电流输出型 D/A 转换器，转换时间 1 μs，片内带输入数字锁存器。DAC0832 与单片机接成数据直接写入方式，当单片机把一个数据写入 DAC 寄存器时，DAC0832 的输出模拟电压信号随之相应变化。利用 D/A 转换器可以产生各种波形，如方波、三角波、锯齿波等以及它们组合产生的复合波形和不规则波形。这些复合波形利用标准的测试设备是很难产生的。

12.4　程序设计

1. 初始化程序

初始化程序的主要工作是设置定时器的工作模式、初值预置、开中断、打开定时器等。在这里定时器 T0 工作于 16 位定时模式，单片机按定时时间重复地把波形数据送到DAC0832 的寄存器。初始化程序流程图见图 12.3。

2. 键扫描程序

键扫描程序的任务是检查 3 个按键是否被按下,如有按下则执行相应的功能。这里 3 个按键分别用于频率增加、频率减小和正弦波与三角波的选择功能,其程序流程图见图 12.4。

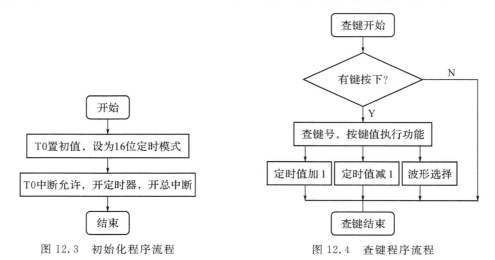

图 12.3　初始化程序流程　　　　　　　　图 12.4　查键程序流程

3. 波形数据产生程序

波形数据产生程序是定时器 T0 的中断程序,当定时器计数溢出时发生一次中断,当发生中断时,单片机将按次序将波形数据表中的波形数据——送入 DAC0832,DAC0832 根据输入的数据大小输出相应的电压,波形数据产生程序流程见图 12.5。

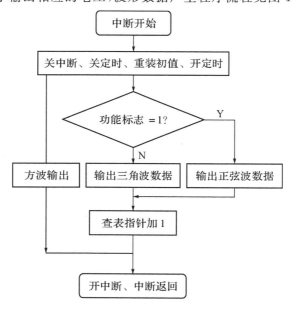

图 12.5　波形数据产生程序流程

4. 主程序

主程序的任务是进行上电初始化,并在程序运行中不断查询按键情况,执行相应的功能。

12.5　软件调试与运行结果

硬件仿真电路的调试较简单,只要元器件连线无误,一般能一次成功。软件的调试主要是各子程序的调试,其中频率的增减按键因为计数器为 16 位定时器,最大值为 65535,加或减 1 变化很慢,可在加减时用 255 作为加减数,使频率的调整变化较快些,但是在接近最高频率时变化太快。如果加减时用 1 作为加减数,那么在频率的高端变化平稳,而在频率的低端则变化太慢。在调试时可根据应用特点选择加减数的大小。

简易低频信号源输出的频率不是很大,在设计中一周期波形用了 256 个采样点合成,波形不是很光滑,如增加采样点,则输出的频率会更低,在设计中应根据应用特点选择合理的采样点数。用单片机产生低频率信号的最大优点是可以输出产生复杂的不规则波形,这是一般的通用信号源无法做到的。图 12.6 及图 12.7 分别为正弦波仿真及三角波仿真输出时的窗口图。

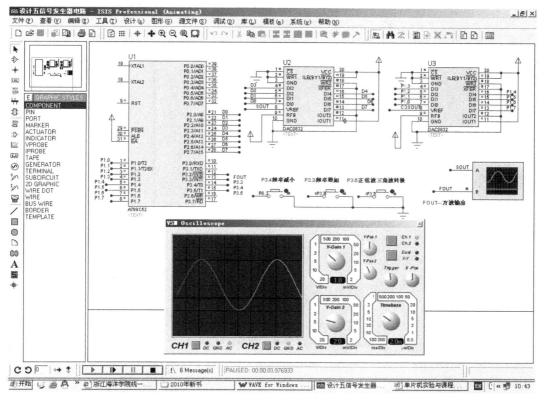

图 12.6　正弦波仿真窗口

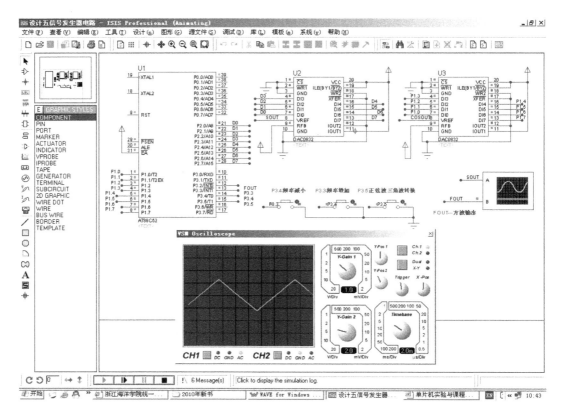

图 12.7   三角波仿真窗口

12.6   源程序清单

12.6.1   课程设计五参考汇编程序

```
;******************************;
;课程设计五程序:低频信号发生器    ;
;   正弦波/三角波发生器           ;
;       12MHz 晶振               ;
;******************************;
;
;正弦波发生器,key0 口按键增加输出频率,key1 口按键减小输出频率
;sinout 口输出正弦波,cosout 口输出余弦波,使用定时器 T0,16 位定时模式。
;R6、R7 用作 10ms 延时寄存器,Fout 输出方波
;
key0      bit      P3.3          ;频率减小按键
```

```
key1        bit     P3.4            ;频率增加按键
key2        bit     P3.5            ;正弦波/三角波转换按键
sinout      EQU     P2              ;正弦波/三角波输出
;cosout     EQU     P1              ;余弦波输出
FOUT        BIT     P3.2            ;方波输出
FLAG        BIT     00H             ;正弦波/三角波输出标志
SINP        DATA    30H             ;正弦波查表指针
COSP        DATA    31H             ;余弦波查表指针
THOD        DATA    32H             ;定时器初值存放(高 8 位)
TLOD        DATA    33H             ;定时器初值存放(低 8 位)
;
ORG         0000H
LJMP        START
ORG         000BH
LJMP        INTT0
;
ORG         0060H
;
START:  MOV     SP,#70H
        MOV     SINP,#00H
        MOV     COSP,#40H
        MOV     TMOD,#11H
        MOV     THOD,#0FFH          ;初值,决定波形频率
        MOV     TLOD,#0AFH
        MOV     TH0,THOD
        MOV     TL0,TLOD
        clr     FLAG                ;
CONZS:  CPL     FLAG                ;0 正弦波,1 三角波
        JB      FLAG,NOSIN
        MOV     DPTR,#LIST
MAIN0:  SETB    ET0
        SETB    EA
        SETB    TR0
MAIN:   JNB     key0,INCKEY
        JNB     key1,DECKEY
        JNB     key2,CONSIN
;       ORL     PCON,#01H
        LJMP    MAIN
;
```

```
NOSIN:  MOV    DPTR,♯LIST1
        AJMP   MAIN0
;按键功能,输出频率增大
INCKEY: LCALL  DL10ms
        JB     key0,MAIN
        MOV    A,TL0D
        CJNE   A,♯0AFH,INC1
        LJMP   MAIN
INC1:   INC    TL0D
        LJMP   MAIN
;按键功能,输出频率减小
DECKEY: LCALL  DL10ms
        JB     key1,MAIN
        MOV    A,TL0D
        CJNE   A,♯00H,DEC1
        LJMP   MAIN
DEC1:   DEC    TL0D
        LJMP   MAIN
;按键功能,输出波形转换
CONSIN: LCALL  DL10ms
        JB     key2,MAIN
WAITOFF:JNB    key2,WAITOFF    ;等待按键释放
        LCALL  DL10ms
        JNB    key2,WAITOFF
        AJMP   CONZS


;定时器 T0 中断程序
INTT0:  PUSH   ACC
        CPL    Fout             ;方波输出,作辅助功能用
        MOV    TH0,TH0D
        MOV    TL0,TL0D
        MOV    A,SINP
        MOVC   A,@A+DPTR
        MOV    sinout,A         ;正弦波从 sinout 口输出
;       MOV    A,COSP
;       MOVC   A,@A+DPTR        ;
;       MOV    cosout,A         ;余弦波从 cosout 口输出
        INC    SINP
;       INC    COSP
        POP    ACC
```

```
                RETI
;10ms 延时程序
DL512：    MOV     R7,♯0FFH
LOOP：     DJNZ    R7,LOOP
           RET
DL10ms：   MOV     R6,♯10
LOOP1：    LCALL   DL512
           DJNZ    R6,LOOP1
           RET
```

;正弦函数表(共 256 个点,每点 1.40625 度)

```
LIST：    DB      80H,83H,85H,88H,8AH,8DH,8FH,92H
          DB      94H,97H,99H,9BH,9EH,0A0H,0A3H,0A5H
          DB      0A7H,0AAH,0ACH,0AEH,0B1H,0B3H,0B5H,0B7H
          DB      0B9H,0BBH,0BDH,0BFH,0C1H,0C3H,0C5H,0C7H
          DB      0C9H,0CBH,0CCH,0CEH,0D0H,0D1H,0D3H,0D4H
          DB      0D6H,0D7H,0D8H,0DAH,0DBH,0DCH,0DDH,0DEH
          DB      0DFH,0E0H,0E1H,0E2H,0E3H,0E3H,0E4H,0E4H
          DB      0E5H,0E5H,0E6H,0E6H,0E7H,0E7H,0E7H,0E7H
          DB      0E7H,0E7H,0E7H,0E7H,0E6H,0E6H,0E5H,0E5H
          DB      0E4H,0E4H,0E3H,0E3H,0E2H,0E1H,0E0H,0DFH
          DB      0DEH,0DDH,0DCH,0DBH,0DAH,0D8H,0D7H,0D6H
          DB      0D4H,0D3H,0D1H,0D0H,0CEH,0CCH,0CBH,0C9H
          DB      0C7H,0C5H,0C3H,0C1H,0BFH,0BDH,0BBH,0B9H
          DB      0B7H,0B5H,0B3H,0B1H,0AEH,0ACH,0AAH,0A7H
          DB      0A5H,0A3H,0A0H,9EH,9BH,99H,97H,94H
          DB      92H,8FH,8DH,8AH,88H,85H,83H,80H
          DB      7DH,7BH,78H,76H,73H,71H,6EH,6CH
          DB      69H,67H,65H,62H,60H,5DH,5BH,59H
          DB      56H,54H,52H,4FH,4DH,4BH,49H,47H
          DB      45H,43H,41H,3FH,3DH,3BH,39H,37H
          DB      35H,34H,32H,30H,2FH,2DH,2CH,2AH
          DB      29H,28H,26H,25H,24H,23H,22H,21H
          DB      20H,1FH,1EH,1DH,1DH,1CH,1CH,1BH
          DB      1BH,1AH,1AH,1AH,19H,19H,19H,19H
          DB      19H,19H,19H,19H,1AH,1AH,1AH,1BH
          DB      1BH,1CH,1CH,1DH,1DH,1EH,1FH,20H
          DB      21H,22H,23H,24H,25H,26H,28H,29H
          DB      2AH,2CH,2DH,2FH,30H,32H,34H,35H
          DB      37H,39H,3BH,3DH,3FH,41H,43H,45H
          DB      47H,49H,4BH,4DH,4FH,52H,54H,56H
```

```
        DB      59H,5BH,5DH,60H,62H,65H,67H,69H
        DB      6CH,6EH,71H,73H,76H,78H,7BH,7DH
;三角波函数表
LIST1： DB      80H,81H,82H,83H,84H,85H,86H,87H
        DB      88H,89H,8AH,8BH,8CH,8DH,8EH,8FH
        DB      90H,91H,92H,93H,94H,95H,96H,97H
        DB      98H,99H,9AH,9BH,9CH,9DH,9EH,9FH
        DB      0A0H,0A1H,0A2H,0A3H,0A4H,0A5H,0A6H,0A7H
        DB      0A8H,0A9H,0AAH,0ABH,0ACH,0ADH,0AEH,0AFH
        DB      0B0H,0B1H,0B2H,0B3H,0B4H,0B5H,0B6H,0B7H
        DB      0B8H,0B9H,0BAH,0BBH,0BCH,0BDH,0BEH,0BFH
        DB      0BFH,0BEH,0BDH,0BCH,0BBH,0BAH,0B9H,0B8H
        DB      0B7H,0B6H,0B5H,0B4H,0B3H,0B2H,0B1H,0B0H
        DB      0AFH,0AEH,0ADH,0ACH,0ABH,0AAH,0A9H,0A8H
        DB      0A7H,0A6H,0A5H,0A4H,0A3H,0A2H,0A1H,0A0H
        DB      9FH,9EH,9DH,9CH,9BH,9AH,99H,98H
        DB      97H,96H,95H,94H,93H,92H,91H,90H
        DB      8FH,8EH,8DH,8CH,8BH,8AH,89H,88H
        DB      87H,86H,85H,84H,83H,82H,81H,80H
        DB      7FH,7EH,7DH,7CH,7BH,7AH,79H,78H
        DB      77H,76H,75H,74H,73H,72H,71H,70H
        DB      6FH,6EH,6DH,6CH,6BH,6AH,69H,68H
        DB      66H,66H,65H,64H,63H,62H,61H,60H
        DB      5FH,5EH,5DH,5CH,5BH,5AH,59H,58H
        DB      55H,55H,55H,54H,53H,52H,51H,50H
        DB      4FH,4EH,4DH,4CH,4BH,4AH,49H,48H
        DB      44H,44H,45H,44H,43H,42H,41H,40H
        DB      40H,41H,42H,43H,44H,45H,46H,47H
        DB      48H,49H,4AH,4BH,4CH,4DH,4EH,4FH
        DB      50H,51H,52H,53H,55H,55H,56H,57H
        DB      58H,59H,5AH,5BH,5CH,5DH,5EH,5FH
        DB      60H,61H,62H,63H,66H,65H,66H,67H
        DB      68H,69H,6AH,6BH,6CH,6DH,6EH,6FH
        DB      70H,71H,72H,73H,77H,75H,76H,77H
        DB      78H,79H,7AH,7BH,7CH,7DH,7EH,7FH
;
        END                     ;结束
```

12.6.2　课程设计五参考 C 程序

```
/*------------------------------------
Wave Generator program V11.1
MCU STC89C52RC   XAL 12MHz
Build by Gavin Hu，2010.6.15
------------------------------------ */
# include <reg51.h>
sbit UPKEY = P3^3；
sbit DOWNKEY = P3^4；
sbit SHKEY = P3^5；
sbit FOUT = P3^2；
void delay_ms(unsigned int)；
unsigned char fun = 0；
unsigned char th0_reg,tl0_reg；
/*------------------------------------
   main function
------------------------------------ */
void main(void)
{
unsigned int f = 500；
TMOD = 0x01；
th0_reg = (unsigned int)(65539.0 - (19531.25/f)) >> 8；
tl0_reg = (unsigned int)(65539.0 - (19531.25/f)) & 0x00ff；
TH0 = th0_reg；
TL0 = tl0_reg；
TR0 = 1；
IE = 0x82；
P3 = 0xff；
while(1)
    {
    if ((UPKEY == 0)&&(f<1000))
        {
        f += (f/10)? (f/10):1；
        th0_reg = (unsigned int)(65539.0 - (19531.25/f)) >> 8；
        tl0_reg = (unsigned int)(65539.0 - (19531.25/f)) & 0x00ff；
        delay_ms(100)；
        }
```

```c
    if ((DOWNKEY == 0)&&(f>1))
        {
        f -= (f/10)? (f/10):1;
        th0_reg = (unsigned int)(65539.0 - (19531.25/f)) >> 8;
        tl0_reg = (unsigned int)(65539.0 - (19531.25/f)) & 0x00ff;
        delay_ms(100);
        }
    if (SHKEY == 0)
        {
        fun = ! fun;
        delay_ms(100);
        }
    }
}


/*------------------------------------
  Delay function
  Parameter: unsigned int dt
  Delay time = dt(ms)
  ------------------------------------ */
void delay_ms(unsigned int dt)
{
register unsigned char bt,ct;
for (;dt;dt --)
    for (ct = 2;ct;ct --)
        for (bt = 248; -- bt;);
}


/*------------------------------------
  T0 interrupt function
  ------------------------------------ */
void intt0(void) interrupt 1
{
unsigned char code sin_table[] = {
0,0,0,0,0,0,0,0,1,1,1,1,1,2,2,2,2,3,3,3,4,4,5,5,6,6,6,7,8,8,9,9,
10,10,11,12,12,13,14,14,15,16,17,17,18,19,20,21,22,23,23,24,25,
26,27,28,29,30,31,32,33,34,35,37,38,39,40,41,42,43,45,46,47,48,
49,51,52,53,54,56,57,58,60,61,62,64,65,66,68,69,71,72,73,75,76,
78,79,81,82,84,85,87,88,90,91,93,94,96,97,99,100,102,103,105,106,
```

```
108,109,111,113,114,116,117,119,120,122,124,125,127,128,130,131,
133,135,136,138,139,141,142,144,146,147,149,150,152,153,155,156,
158,159,161,162,164,165,167,168,170,171,173,174,176,177,179,180,
182,183,184,186,187,189,190,191,193,194,195,197,198,199,201,202,
203,204,206,207,208,209,210,212,213,214,215,216,217,218,220,221,
222,223,224,225,226,227,228,229,230,231,232,232,233,234,235,236,
237,238,238,239,240,241,241,242,243,243,244,245,245,246,246,247,
247,248,249,249,249,250,250,251,251,252,252,252,253,253,253,253,
254,254,254,254,254,255,255,255,255,255,255,255,255};
static unsigned char wave_index = 0;
static bit index_sign = 1;
TL0 = tl0_reg;
TH0 = th0_reg;
if (fun) {P2 = sin_table[wave_index];}
    else {P2 = wave_index;}
FOUT = ! FOUT;
if (index_sign == 1) {if ( ++ wave_index == 255) index_sign = 0;}
    else {if (-- wave_index == 0) index_sign = 1;}
}
```

第 13 章

课程设计六:16 点阵 LED 显示器

13.1　系统功能

设计一个能显示 4 个 16×16 点阵中文汉字或图形的 LED 显示屏,要求有静止、上移等显示方式。

13.2　设计方案

点阵 LED 图文显示的原理是通过控制一些组成某些图形或文字的各个点所在位置相对应的 LED 器件是否发光,然后得到我们想要看到的显示结果,这种同时控制多个发光点亮灭的方法称为静态驱动显示方式。每个 16×16 的点阵文字共有 256 个发光二极管,显然单片机没有这么多端口,4 个文字的话更要有 1024 个控制端口。如果采用锁存器来扩展端口,按 8 位的锁存器来计算,1 个 16×16 的点阵需要 256/8＝32 个锁存器。这个数字很庞大,在实际应用中的显示屏往往还要大得多,这样在锁存器上花的成本将是一个很庞大的数字。因此,在实际应用中的显示屏采用另一种称为动态扫描的显示方法。

所谓动态扫描,简单地说就是逐行轮流点亮,这样扫描驱动电路就可以实现多行(比如 16 行)的同名列共用一套列驱动器。就 16×16 的点阵来说,把所有同一行的发光管的阳极连在一起,把所有同一列的发光管的阴极连在一起(共阳的接法),先送出对应第一行发光管亮灭的数据并锁存,然后选通第一行使其点亮一定的时间,然后熄灭;再送出第二行的数据并锁存,然后选通第二行使其点亮相同的时间,然后熄灭;⋯⋯ 第十六行之后又重新点亮第一行,这样反复轮回。当轮回的速度足够快(每秒 24 次以上),由于人眼的视觉暂留现象,我们就能看到显示屏上稳定的图形了。

采用扫描方式进行显示时,每行有一个行驱动器,各行的同名列共用一个列驱动器。

显示数据通常存储在单片机的存储器中，按 8 位一个字节的形式顺序排放。显示时要把一行中各列的数据都传送到相应的列驱动器上去，这就存在一个显示数据传输的问题。从控制电路到列驱动器的数据传输可以采用并行方式或串行方式。显然，采用并行方式时，从控制电路到列驱动器的线路数量大，相应的硬件数目多。当列数很多时，并行传输的方案是不可取的。

采用串行传输的方法，控制电路可以只用一根信号线，将列数据一位一位传往列驱动器，在硬件方面无疑是十分经济的。但是，串行传输过程较长，数据按顺序一位一位地输出给列驱动器，只有当一行的各列数据都已传输到位之后，这一行的各列才能并行地进行显示。这样，对于一行的显示过程就可以分解成列数据准备（传输）和列数据显示两个部分。对于串行传输方式来说，列数据准备时间相对要长一些，在行扫描周期确定的情况下，行显示的时间就会减小一点（比如四个水平排列的 16×16 文字的 LED 每行亮 1ms，串行传送时间会花约 0.1ms），水平文字较多时会影响到 LED 的亮度效果。

解决串行传输中列数据准备和列数据显示的时间矛盾问题，可以采用重叠处理的方法来解决。即在显示本行各列数据的同时，传送下一行的列数据。为了达到重叠处理的目的，列数据的显示就需要具有锁存功能。经过上述分析，可知列驱动器电路应具备能实现串入/并出的移位功能，同时具有并行锁存输出的功能。这样，在本行已准备好的数据输入并行锁存器输出端口进行显示时，串并移位寄存器就可以准备下一行的列数据传送，而不会影响每行的显示时间。图 13.1 为 16 点阵 LED 显示器系统结构框图。

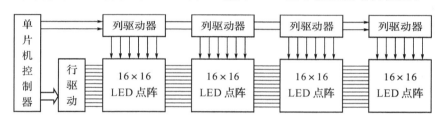

图 13.1　16 点阵 LED 显示器系统结构

13.3　系统硬件仿真电路

系统硬件电路大致上可以分成单片机系统与外围电路、行驱动电路和列驱动电路三部分。

13.3.1　单片机系统与外围电路

单片机采用 89C52 系列，采用 12MHz 或以上更高频率的晶振，以获得较高的刷新频率，使显示更稳定。单片机的串口与列驱动器相连，用来传送显示数据。P1 口低 4 位与行驱动器相连（4/16 译码器），送出行选信号；P1.5～P1.7 口则用来发送移位控制信号。系统硬件仿真电路见图 13.2。

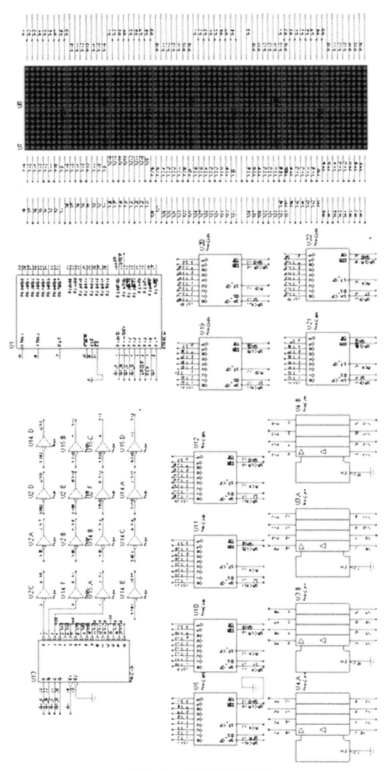

图 13.2　系统硬件仿真电路

13.3.2　行驱动电路

单片机 P1 口低 4 位输出的行号经 4/16 线译码器 74LS154 译码后生成 16 条行选通信号线，再经过驱动器驱动对应的行线。一条行线上要带动 16 列×4 的 LED 进行显示，按每一 LED 器件 5mA 电流计算，64 个 LED 同时发光时，需要 320mA 电流，仿真电路中选用反相器并经 74HC244 驱动四个 LED 块的行扫描供电。

13.3.3　列驱动电路

列驱动电路由集成电路 74HC595 构成，它具有一个串入/并出（8 位）的移位寄存器和一个 8 位并行输出锁存器的结构，而且移位寄存器和输出锁存器的控制是各自独立的，可以实现在显示本行各列数据的同时，串行传送下一行的列数据，即达到时间上重叠处理的能力。

13.4　程序设计

显示屏软件的主要功能是向屏幕提供显示数据，并产生各种控制信号，使屏幕按设计的要求显示。根据软件分层次设计的原理，我们可把显示屏的软件系统分成两大层：第一层是底层的显示驱动程序，第二层是上层的系统应用程序。显示驱动程序负责向屏体传送显示数据，并负责产生行扫描信号和其他控制信号，配合完成 LED 显示屏的扫描显示工作。显示驱动程序由定时器 T0 中断程序实现。系统应用程序完成系统环境设置（初始化）、显示效果处理等工作，由主程序来实现。

13.4.1　显示驱动程序

显示驱动程序查询当前点亮的行号，从显示缓存区内读取下一行的显示数据，并通过串口发送给移位寄存器。为消除在切换行显示数据的时候产生拖尾现象，驱动程序先要关闭显示屏，即消隐，等显示数据输入输出锁存器并锁存，然后再输出新的行号，重新打开显示。图 13.3 为显示驱动程序（显示屏扫描程序）流程图。

13.4.2　系统主程序

系统主程序首先是对系统运行环境初始化，包括设置串口、定时器、中断和端口。然后以"静止翻屏"效果显示文字或图案，每次 4 个文字，停留约几秒钟后接着显示后面的文字，一遍文字显示完毕后，接着进行向上连续的滚动显示汉字。显示效果可以根据需要进行设置,系统程序会不断地循环执行显示效果。图 13.4 是系统主程序的流程图。

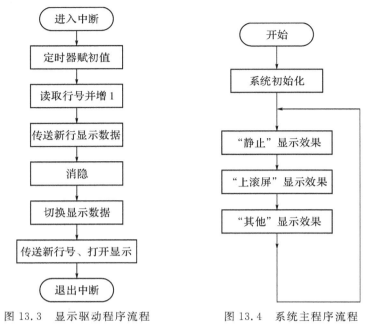

图 13.3　显示驱动程序流程　　　　　图 13.4　系统主程序流程

13.5　软件调试与运行结果

LED 显示器硬件仿真电路只要器件连线可靠,一般无需调试即可正常工作。软件部分需要调试的主要有显示屏刷新频率及显示效果两部分。显示屏刷新率由定时器 T0 的溢出率和单片机的晶振频率决定。从理论上来说,24Hz 以上的刷新率就能看到连续稳定的显示,刷新率越高,显示越稳定,同时刷新率越高,显示驱动程序占用的 CPU 时间也越多。实验证明,在目测条件下刷新率 40Hz 以下的画面看起来闪烁较严重,刷新率50Hz 以上的已基本觉察不出画面闪烁,刷新率达到 85Hz 以上时,画面闪烁将没有明显改善。表 13.1 为采用 24MHz 晶振时点阵显示屏刷新频率及其对应的定时器 T0 初值。

表 13.1　显示屏刷新率(帧频)与 T0 初值关系表(24MHz 晶振时)

刷新率(Hz)	25	50	62.5	75	85	100	120
T0 初值	0xec78	0xf63c	0xf830	0xf97e	0xfa42	0xfb1e	0xfbee

图 13.5 为用 Proteus 进行程序仿真运行的效果图。

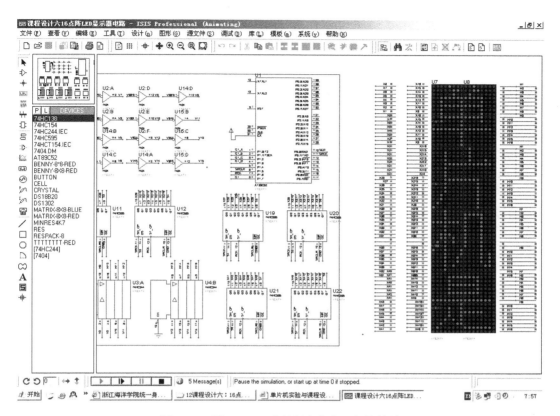

图 13.5   用 Proteus 进行程序仿真运行的效果

13.6   源程序清单

13.6.1   课程设计六参考汇编程序

```
;**************************************;
;              课程设计六程序            ;
;      四字 16×16 点阵电子屏字符显示器      ;
;           89C52   12MHz 晶振            ;
;                                      ;
;**************************************;
```

;显示字用查表法，不占内存，用 4 个 16×16 共阳 LED 点阵显示

;效果：向上滚动显示字，每次 4 个，重复循环

;R2：行扫描地址（从 00～0FH）

;R3：滚动显示时控制移动速度，也可控制静止显示的时间

```
;************;
;中断入口程序;
;************;
;
ORG     0000H
LJMP    START
;
ORG     000BH
LJMP    INTT0
;
;************;
;   主程序    ;
;************;
;
START:   MOV    20H,#00H        ;清标志,00H为1帧扫描结束标志
         MOV    A,#0FFH         ;端口初始化
         MOV    P1,A
         MOV    P2,A
         MOV    P3,A
         MOV    P0,A
         CLR    P1.6            ;串行寄存器输入输出端控制位
         MOV    TMOD,#01H       ;使用T0作16位定时器,行扫描用
         MOV    TH0,#0FCH       ;1ms初值(12MHz)
         MOV    TL0,#18H
         MOV    SCON,#00H       ;串口0方式传送显示字节
         MOV    IE,#82H         ;T0中断允许,总中断允许
         MOV    SP,#70H
         LCALL  DIS1            ;显示准备,黑屏,1.5秒
MAIN:    MOV    DPTR,#TAB
         LCALL  MOVDISP         ;逐排显示,每次4字
         MOV    DPTR,#TAB
         LCALL  MOVDISP1        ;滚动显示,每排4字
         AJMP   MAIN
;
;
;********************;
;四字逐排显示子程序   ;
;********************;
;每次四字移入移出显示方式,入口时定义好DPTR值
```

```
;
MOVDISP:   MOV     R1,♯6           ;显示 6 排字,每排 4 字(R1 = 排数)
DISLOOP:   MOV     R3,♯100         ;每排显示时间 16ms×100 = 1.6s
DISMOV:    MOV     R2,♯00H         ;第 0 行开始
           SETB    TR0             ;开始扫描(每次一帧)
WAITMOV:   JBC     00H,DISMOV1     ;标志为 1 扫描一帧结束(16ms 为 1 帧,每行 1ms)
           AJMP    WAITMOV
DISMOV1:   DJNZ    R3,DISMOV       ;1 帧重复显示(控制显示时间)
           MOV     A,♯128          ;显示字指针移一排(每排 4 字×32 = 128 字)
           ADD     A,DPL           ;
           MOV     DPL,A
           MOV     A,♯0
           ADDC    A,DPH
           MOV     DPH,A
           DEC     R1              ;R1 为 0,显示完
           MOV     A,R1
           JZ      MOVOUT          ;
           AJMP    DISLOOP         ;
MOVOUT:    RET                     ;移动显示结束
;
;******************** ;
;   四字滚动显示子程序 ;
;******************** ;
;每排 4 字向上移出显示方式,入口时定义好 DPTR 值
;
MOVDISP1:  MOV     R1,♯255         ;向上移动显示 6 排字,每排 4 字(R1 = 排数 * 16)
DISLOOP1:  MOV     R3,♯10          ;移动速度 16ms×10 = 0.16s
DISMOV2:   MOV     R2,♯00H         ;第 0 行开始
           SETB    TR0             ;开始扫描(每次一帧)
WAITMOV1:  JBC     00H,DISMOV3     ;标志为 1 扫描一帧结束(16ms 为 1 帧,每行 1ms)
           AJMP    WAITMOV1
DISMOV3:   DJNZ    R3,DISMOV2      ;1 帧重复显示(控制移动速度)
           INC     DPTR            ;显示字指针移一行(2 字节位置)
           INC     DPTR
           DEC     R1              ;R1 为 0,显示完
           MOV     A,R1
           JZ      MOVOUT1         ;
           AJMP    DISLOOP1        ;
MOVOUT1:   RET                     ;移动显示结束
```

```
;
;****************** ;
; 四个字显示子程序 ;
;****************** ;
;静止显示表中某四个字
DIS1：    MOV    R3,＃5AH         ;静止显示时间控制(16ms×100＝1.6s)
DIS11：   MOV    R2,＃00H         ;一帧扫描初始值(行地址从 00～0FH)
          MOV    DPTR,＃TAB       ;取表首址
          SETB   TR0             ;开始扫描(每次一帧)
WAIT11：  JBC    00H,DIS111      ;为 1,扫描一帧结束
          AJMP   WAIT11
DIS111：  DJNZ   R3,DIS11
          RET
;
;************ ;
;   扫描程序    ;
;************ ;
;1ms 传送一行,每行显示 1ms,一次传送 4 个字的某行共 8 个字节
;
INTT0：   PUSH   ACC
          MOV    TH0,＃0FCH       ;1ms 初值重装
          MOV    TL0,＃18H
          MOV    A,＃97           ;指向第 4 个字行右字节
          ADD    A,DPL
          MOV    DPL,A
          MOV    A,＃0
          ADDC   A,DPH
          MOV    DPH,A
          MOV    A,＃0
          MOVC   A,@A+DPTR       ;查表
          MOV    SBUF,A          ;串口 0 方式发送
WAIT：    JBC    TI,GO           ;等待发送完毕
          AJMP   WAIT            ;
GO：      MOV    A,DPL           ;指向第 4 个字行左字节
          SUBB   A,＃1
          MOV    DPL,A
          MOV    A,DPH
          SUBB   A,＃0
          MOV    DPH,A
```

```
           MOV     A,♯0
           MOVC    A,@A+DPTR
           MOV     SBUF,A
WAIT1：    JBC     TI,GO1
           AJMP    WAIT1
;
GO1：      MOV     R0,♯03H
MLOOP：    MOV     A,DPL          ;指向前 3 个字行右字节
           SUBB    A,♯31
           MOV     DPL,A
           MOV     A,DPH
           SUBB    A,♯0
           MOV     DPH,A
           MOV     A,♯0
           MOVC    A,@A+DPTR      ;查表
           MOV     SBUF,A         ;串口 0 方式发送
WAIT2：    JBC     TI,GO2         ;等待发送完毕
           AJMP    WAIT2          ;
GO2：      MOV     A,DPL          ;指向前 3 个字行左字节
           SUBB    A,♯1
           MOV     DPL,A
           MOV     A,DPH
           SUBB    A,♯0
           MOV     DPH,A
           MOV     A,♯0
           MOVC    A,@A+DPTR
           MOV     SBUF,A
WAIT3：    JBC     TI,GO3
           AJMP    WAIT3
GO3：      DJNZ    R0,MLOOP       ;执行 3 次
;
           SETB    P1.7           ;关行显示,准备刷新
           NOP                    ;串口寄存器数据稳定
           SETB    P1.6           ;产生上升沿,行数据输入输出端
           NOP                    ;
           NOP                    ;
           CLR     P1.6           ;恢复低电平
           MOV     A,R2           ;修改显示行地址
           ORL     A,♯0F0H        ;修改显示行地址
```

```
        MOV     R2,A            ;修改显示行地址
        MOV     A,P1            ;修改显示行地址
        ORL     A,♯0FH          ;修改显示行地址
        ANL     A,R2            ;修改显示行地址
        MOV     P1,A            ;修改完成
        CLR     P1.7            ;打开行显示
        INC     R2              ;下一行扫描地址值
        INC     DPTR            ;
        INC     DPTR            ;下一行数据地址
        MOV     A,R2
        ANL     A,♯0FH
        JNZ     GO4
        SETB    00H             ;R2 为 10H,现为末行扫描,置 1 帧结束标志
        MOV     A,DPL           ;指针修正为原帧初值
        SUBB    A,♯32
        MOV     DPL,A
        MOV     A,DPH
        SUBB    A,♯0
        MOV     DPH,A
        CLR     TR0             ;一帧扫描完,关扫描
GO4:    POP     ACC
        RETI                    ;退出
;


;***************;
;   扫描文字表    ;
;***************;
;共五排字,每排四个字,前后为黑屏
TAB:
DB  0FFH,0FFH,0FFH,0FFH,0FFH,0FFH,0FFH,0FFH,0FFH,0FFH,0FFH,0FFH,0FFH,0FFH,0FFH,0FFH   ;黑屏
DB  0FFH,0FFH,0FFH,0FFH,0FFH,0FFH,0FFH,0FFH,0FFH,0FFH,0FFH,0FFH,0FFH,0FFH,0FFH,0FFH
DB  0FFH,0FFH,0FFH,0FFH,0FFH,0FFH,0FFH,0FFH,0FFH,0FFH,0FFH,0FFH,0FFH,0FFH,0FFH,0FFH   ;黑屏
DB  0FFH,0FFH,0FFH,0FFH,0FFH,0FFH,0FFH,0FFH,0FFH,0FFH,0FFH,0FFH,0FFH,0FFH,0FFH,0FFH
DB  0FFH,0FFH,0FFH,0FFH,0FFH,0FFH,0FFH,0FFH,0FFH,0FFH,0FFH,0FFH,0FFH,0FFH,0FFH,0FFH   ;黑屏
DB  0FFH,0FFH,0FFH,0FFH,0FFH,0FFH,0FFH,0FFH,0FFH,0FFH,0FFH,0FFH,0FFH,0FFH,0FFH,0FFH
DB  0FFH,0FFH,0FFH,0FFH,0FFH,0FFH,0FFH,0FFH,0FFH,0FFH,0FFH,0FFH,0FFH,0FFH,0FFH,0FFH   ;黑屏
DB  0FFH,0FFH,0FFH,0FFH,0FFH,0FFH,0FFH,0FFH,0FFH,0FFH,0FFH,0FFH,0FFH,0FFH,0FFH,0FFH
DB  0F7H,0DFH,0F9H,0CFH,0BFH,0BFH,0C0H,007H,0DEH,0F7H,0C0H,007H,0DEH,0F7H,0DEH,0F7H   ;单
DB  0C0H,007H,0DEH,0F7H,0FEH,0FFH,000H,001H,0FEH,0FFH,0FEH,0FFH,0FEH,0FFH,0FEH,0FFH
```

```
    DB    0FFH,0BFH,0EFH,0BFH,0EFH,0BFH,0EFH,0BBH,0E0H,001H,0EFH,0FFH,0EFH,0FFH,0EFH,0FFH    ;片
    DB    0E0H,00FH,0EFH,0EFH,0EFH,0EFH,0EFH,0EFH,0DFH,0EFH,0DFH,0EFH,0BFH,0EFH,07FH,0EFH
    DB    0EFH,0FFH,0EFH,007H,0EFH,077H,001H,077H,0EFH,077H,0EFH,077H,0C7H,077H,0CBH,077H    ;机
    DB    0ABH,077H,0AFH,077H,06EH,0F7H,0EEH,0F5H,0EDH,0F5H,0EDH,0F5H,0EBH,0F9H,0EFH,0FFH
    DB    0FFH,0FFH,0F0H,00FH,0F7H,0EFH,0F0H,00FH,0F7H,0EFH,0F0H,00FH,0FFH,0FFH,000H,001H    ;是
    DB    0FEH,0FFH,0F6H,0FFH,0F6H,007H,0F6H,0FFH,0EAH,0FFH,0DCH,0FFH,0BFH,001H,0FFH,0FFH
    DB    0FFH,0FFH,0C0H,003H,0FEH,0FFH,0FEH,0FFH,0FEH,0FFH,0FEH,0FFH,0FEH,0FFH,0FEH,0FFH    ;工
    DB    0FEH,0FFH,0FEH,0FFH,0FEH,0FFH,0FEH,0FFH,0FEH,0FFH,000H,001H,0FFH,0FFH,0FFH,0FFH
    DB    0FBH,0BFH,0FBH,0BFH,0FBH,0BFH,0FBH,0BBH,0BBH,0B9H,0DBH,0B3H,0DBH,0B7H,0EBH,0AFH    ;业
    DB    0E3H,0AFH,0EBH,09FH,0FBH,0BFH,0FBH,0BFH,0FBH,0BBH,000H,001H,0FFH,0FFH,0FFH,0FFH
    DB    0FEH,0FFH,0FEH,0FFH,0DEH,0F7H,0C0H,003H,0DEH,0F7H,0DEH,0F7H,0DEH,0F7H,0DEH,0F7H    ;中
    DB    0DEH,0F7H,0C0H,007H,0DEH,0F7H,0FEH,0FFH,0FEH,0FFH,0FEH,0FFH,0FEH,0FFH,0FEH,0FFH
    DB    0E0H,00FH,0EFH,0EFH,0E0H,00FH,0EFH,0EFH,0E0H,00FH,0FFH,0FFH,000H,001H,0DDH,0FFH    ;最
    DB    0C1H,003H,0DDH,077H,0C1H,0AFH,0DCH,0DFH,0C1H,0AFH,01DH,071H,0FCH,0FBH,0FDH,0FFH    ;
    DB    0F7H,0DFH,0F7H,0DFH,080H,003H,0F7H,0DFH,0F0H,01FH,0F7H,0DFH,0F0H,01FH,0F7H,0DFH    ;基
    DB    000H,001H,0F7H,0DFH,0EEH,0E7H,0C0H,011H,03EH,0FBH,0FEH,0FFH,080H,003H,0FFH,0FFH
    DB    0FEH,0FFH,0FEH,0FFH,0FEH,0FFH,000H,001H,0FCH,07FH,0FCH,0BFH,0FAH,0BFH,0FAH,0DFH    ;本
    DB    0F6H,0EFH,0EEH,0E7H,0D0H,011H,03EH,0FBH,0FEH,0FFH,0FEH,0FFH,0FEH,0FFH,0FFH,0FFH
    DB    0EFH,07FH,0EFH,07FH,0DFH,07FH,083H,003H,0BAH,0FBH,0BAH,0FBH,0B9H,0FBH,083H,07BH    ;的
    DB    0BBH,0BBH,0BBH,09BH,0BBH,0DBH,0BBH,0FBH,083H,0FBH,0BBH,0D7H,0BFH,0EFH,0FFH,0FFH
    DB    0FEH,0FFH,0FFH,07FH,0C0H,003H,0DFH,0FFH,0DDH,0FFH,0DEH,0F7H,0CFH,073H,0D7H,037H    ;应
    DB    0DBH,06FH,0DBH,06FH,0D9H,0DFH,0BBH,0DFH,0BFH,0BFH,0A0H,001H,07FH,0FFH,0FFH,0FFH    ;
    DB    0FFH,0FFH,0E0H,003H,0EFH,07BH,0EFH,07BH,0EFH,07BH,0E0H,003H,0EFH,07BH,0EFH,07BH    ;用
    DB    0EFH,07BH,0E0H,003H,0EFH,07BH,0EFH,07BH,0DFH,07BH,0DFH,07BH,0BFH,06BH,07FH,077H
    DB    0FDH,0FFH,0FEH,0FFH,0FFH,07FH,000H,001H,0FDH,0FFH,0FDH,0FFH,0FCH,00FH,0FDH,0EFH    ;方
    DB    0FBH,0EFH,0FBH,0EFH,0F7H,0EFH,0F7H,0EFH,0EFH,0DFH,06FH,03FH,09FH,0FFH,0FFH
    DB    0FFH,05FH,0FFH,067H,0FFH,06FH,080H,003H,0FFH,07FH,0FFH,07FH,0FFH,07FH,0C1H,07FH    ;式
    DB    0F7H,0BFH,0F7H,0BFH,0F7H,0BFH,0F4H,0DFH,0E3H,0DDH,08FH,0EDH,0DFH,0F5H,0FFH,0FBH
    DB    0F9H,0BFH,0C7H,0AFH,0F7H,0B7H,0F7H,0B7H,0F7H,0BFH,000H,001H,0F7H,0BFH,0F7H,0B7H    ;我
    DB    0F1H,0D7H,0C7H,0CFH,037H,0DFH,0F7H,0AFH,0F6H,06DH,0F7H,0F5H,0D7H,0F9H,0EFH,0FDH
    DB    0FFH,007H,0C0H,06FH,0EDH,0EFH,0F6H,0DFH,0C0H,001H,0DDH,0FDH,0BDH,0FFH,0C0H,003H    ;爱
    DB    0FBH,0FFH,0F8H,00FH,0F3H,0DFH,0F4H,0BFH,0EFH,03FH,09CH,0CFH,073H,0F1H,0CFH,0FBH
    DB    0F7H,0DFH,0F9H,0CFH,0FBH,0BFH,0C0H,007H,0DEH,0F7H,0C0H,007H,0DEH,0F7H,0DEH,0F7H    ;单
    DB    0C0H,007H,0DEH,0F7H,0FEH,0FFH,000H,001H,0FEH,0FFH,0FEH,0FFH,0FEH,0FFH,0FEH,0FFH
    DB    0FFH,0BFH,0EFH,0BFH,0EFH,0BFH,0EFH,0BBH,0E0H,001H,0EFH,0FFH,0EFH,0FFH,0EFH,0FFH    ;片
    DB    0E0H,00FH,0EFH,0EFH,0EFH,0EFH,0EFH,0EFH,0DFH,0EFH,0DFH,0EFH,0BFH,0EFH,07FH,0EFH
    DB    0EFH,0FFH,0EFH,007H,0EFH,077H,001H,077H,0EFH,077H,0EFH,077H,0C7H,077H,0CBH,077H    ;机
    DB    0ABH,077H,0AFH,077H,06EH,0F7H,0EEH,0F5H,0EDH,0F5H,0EDH,0F5H,0EBH,0F9H,0EFH,0FFH    ;
    END              ;结束
```

13.6.2　课程设计六参考 C 程序

```
/*----------------------------------------
16×64 点阵 LED 显示屏程序    V12.1
MCU AT89C52    XAL 24MHz
Build by Gavin Hu, 2010.6.15
---------------------------------------- */
# include "reg52.h"
# define BLKNUM 4
# define BREGNUM 32
sbit G = P1^7;                              //P1.7 为显示允许控制信号端口
sbit R_CLK = P1^6;                         //P1.6 为输出锁存器时钟信号端
sbit S_CLR = P1^5;                         //P1.5 为移位寄存器清零端
void delay_ms(unsigned int);               //延时函数
unsigned char idata dispram[BLKNUM][BREGNUM];   //显示区缓存


/*----------------------------------------
   主函数 void main(void)
---------------------------------------- */
void main(void)
{
# define CHARNUM 12
unsigned char code Bmp[][BREGNUM] = {\
/*-- 文字：  单   --*/
/*--  宋体 12；此字体下对应的点阵为：宽×高 = 16×16   -- */
0xF7,0xDF,0xF9,0xCF,0xFB,0xBF,0xC0,0x07,0xDE,0xF7,0xC0,0x07,0xDE,0xF7,0xDE,
0xF7,0xC0,0x07,0xDE,0xF7,0xFE,0xFF,0x00,0x01,0xFE,0xFF,0xFE,0xFF,0xFE,0xFF,
0xFE,0xFF,


/*--  文字：  片   --*/
/*--  宋体 12；  此字体下对应的点阵为：宽×高 = 16×16    -- */
0xFF,0xBF,0xEF,0xBF,0xEF,0xBF,0xEF,0xBB,0xE0,0x01,0xEF,0xFF,0xEF,0xFF,0xEF,
0xFF,0xE0,0x0F,0xEF,0xEF,0xEF,0xEF,0xEF,0xEF,0xDF,0xEF,0xDF,0xEF,0xBF,0xEF,
0x7F,0xEF,


/*--  文字：  机   --*/
/*--  宋体 12；  此字体下对应的点阵为：宽×高 = 16×16    -- */
0xEF,0xFF,0xEF,0x07,0xEF,0x77,0x01,0x77,0xEF,0x77,0xEF,0x77,0xC7,0x77,0xCB,
```

0x77,0xAB,0x77,0xAF,0x77,0x6E,0xF7,0xEE,0xF5,0xED,0xF5,0xED,0xF5,0xEB,0xF9,
0xEF,0xFF,

/*--　文字：　文　--*/
/*--　宋体 12；　此字体下对应的点阵为：宽×高 = 16×16　--*/
0xFD,0xFF,0xFE,0xFF,0xFE,0xFF,0x00,0x01,0xF7,0xDF,0xF7,0xDF,0xF7,0xDF,0xFB,
0xBF,0xFB,0xBF,0xFD,0x7F,0xFE,0xFF,0xFD,0x7F,0xFB,0x9F,0xE7,0xE1,0x1F,0xF7,
0xFF,0xFF,

/*--　文字：　本　--*/
/*--　宋体 12；　此字体下对应的点阵为：宽×高 = 16×16　--*/
0xFE,0xFF,0xFE,0xFF,0xFE,0xFF,0x00,0x01,0xFC,0x7F,0xFC,0xBF,0xFA,0xBF,0xFA,
0xDF,0xF6,0xEF,0xEE,0xE7,0xD0,0x11,0x3E,0xFB,0xFE,0xFF,0xFE,0xFF,0xFE,0xFF,
0xFF,0xFF,

/*--　文字：　屏　--*/
/*--　宋体 12；　此字体下对应的点阵为：宽×高 = 16×16　--*/
0xC0,0x03,0xDF,0xFB,0xDF,0xFB,0xC0,0x03,0xDB,0xEF,0xDD,0xDF,0xD0,0x03,0xDD,
0xDF,0xDD,0xDF,0xC0,0x01,0xDD,0xDF,0xDD,0xDF,0xBB,0xDF,0xBB,0xDF,0x77,0xDF,
0xEF,0xDF,

/*--　文字：　动　--*/
/*--　宋体 12；　此字体下对应的点阵为：宽×高 = 16×16　--*/
0xFF,0xDF,0xFF,0xDF,0x81,0xDF,0xFF,0xDF,0xFF,0x03,0x00,0xDB,0xEF,0xDB,0xEF,
0xDB,0xDB,0xDB,0xDD,0xDB,0xB0,0xBB,0x05,0xBB,0xBF,0x7B,0xFE,0xEB,0xFD,0xF7,
0xFF,0xFF,

/*--　文字：　态　--*/
/*--　宋体 12；　此字体下对应的点阵为：宽×高 = 16×16　--*/
0xFE,0xFF,0xFE,0xFF,0x80,0x03,0xFE,0xFF,0xFD,0x7F,0xFD,0xBF,0xFA,0xDF,0xF7,
0x67,0xCF,0xF9,0xFE,0xFF,0xFB,0x77,0xDB,0x7B,0xDB,0xED,0x9B,0xED,0xBC,0x0F,
0xFF,0xFF,

/*--　文字：　效　--*/
/*--　宋体 12；　此字体下对应的点阵为：宽×高 = 16×16　--*/
0xEF,0xDF,0xF7,0xCF,0xF7,0xDF,0x80,0xDF,0xEB,0x81,0xED,0xBB,0xDE,0x3B,0xDD,
0xBB,0x9D,0xD7,0xEB,0xD7,0xF7,0xEF,0xF3,0xEF,0xED,0xD7,0xDF,0x31,0x3C,0xFB,
0xFF,0xFF,

```
/*--   文字：  果   --*/
/*--   宋体 12；  此字体下对应的点阵为:宽×高 = 16×16   --*/
0xFF,0xFF,0xE0,0x0F,0xEE,0xEF,0xE0,0x0F,0xEE,0xEF,0xE0,0x0F,0xFE,0xFF,0xFE,
0xFF,0x00,0x01,0xFC,0x7F,0xFA,0xBF,0xF6,0xCF,0xCE,0xF1,0x3E,0xFB,0xFE,0xFF,
0xFE,0xFF,


/*--   文字：  演   --*/
/*--   宋体 12；  此字体下对应的点阵为:宽×高 = 16×16   --*/
0xBF,0x7F,0xDF,0xBF,0xD8,0x01,0xFB,0xFB,0x7C,0x07,0xAF,0xBF,0xEC,0x07,0xDD,
0xB7,0xDC,0x07,0xDD,0xB7,0x3D,0xB7,0xBC,0x07,0xBF,0xFF,0xBE,0xEF,0xBE,0xF7,
0xBD,0xF7,


/*--   文字：  示   --*/
/*--   宋体 12；  此字体下对应的点阵为:宽×高 = 16×16   --*/
0xFF,0xFF,0xE0,0x07,0xFF,0xFF,0xFF,0xFF,0xFF,0x80,0x01,0xFE,0xFF,0xFE,
0xFF,0xEE,0xDF,0xEE,0xEF,0xDE,0xF7,0xBE,0xF3,0x7E,0xFB,0xFE,0xFF,0xFA,0xFF,
0xFD,0xFF
};
unsigned char i,j,k,l;

P1 = 0xbf;                            //P1 端口初值:允许接收、锁存、显示

SCON = 0x00;                          //串口工作模式 0:移位寄存器方式

TMOD = 0x01;                          //定时器 T0 工作方式 1：16 位方式
for (i = 0;i<BLKNUM;i ++ )
    for (j = 0;j<BREGNUM;j ++ )
        dispram[i][j] = 0xff;
TR0 = 1;                              //启动定时器 T0
IE = 0x82;                            //允许定时器 T0 中断
while (1)
    {
    for (i = 0;i<BLKNUM;i ++ )        //显示效果:卷帘入┐
        for (j = 0;j<BREGNUM;j ++ )
            {
            dispram[i][j] = 0;
            if (j%2) delay_ms(100);
            }                         //—————————┘
    //delay_ms(2000);               //延时 2 秒
```

```
        for（i = 0；i＜BLKNUM；i ++）          //显示效果：卷帘出┐
            for（j = 0；j＜BREGNUM；j ++）
                {
                    dispram[i][j] = Bmp[i][j]；
                    if（j % 2）delay_ms(100)；
                }                              //——————┘
        for（l = BLKNUM；l＜CHARNUM；l ++）    //显示效果：上滚屏┐
            for（k = 0；k＜32；k += 2）
                {
                for（i = 0；i＜(BLKNUM − 1)；i ++）
                    {
                    for（j = 0；j＜(BREGNUM − 2)；j ++）
                        {
                            dispram[i][j] = dispram[i][j + 2]；
                        }
                    dispram[i][j] = dispram[i + 1][0]；
                    dispram[i][j + 1] = dispram[i + 1][1]；
                    }
                for（j = 0；j＜(BREGNUM − 2)；j ++）
                    {
                        dispram[i][j] = dispram[i][j + 2]；
                    }
                dispram[i][j] = Bmp[l][k]；
                dispram[i][j + 1] = Bmp[l][k + 1]；
                delay_ms(100)；
                }                              //——————┘
        delay_ms(3000)；                       //延时 3s
        }                                      //end while (1)
}


/*————————————————————————————————
   Delay function
   Parameter：unsigned int dt
   Delay time = dt(ms)
   ———————————————————————————————— */
void delay_ms(unsigned int dt)
{
register unsigned char bt,ct；
for（；dt；dt --）
```

```
    for (ct = 2;ct;ct --)
        for (bt = 248; -- bt;);
}


/* 显示屏扫描（定时器 T0 中断）函数 */
void leddisplay(void) interrupt 1 using 1
{
register unsigned char i,j = BLKNUM;
TH0 = 0xF8;                              //设定显示屏刷新率每秒 62.5 帧
TL0 = 0x30;
i = P1;                                  //读取当前显示的行号
i = ++ i & 0x0f;                         //行号加 1,屏蔽高 4 位
do {
    j -- ;
    TI = 0;
    SBUF = dispram[j][i * 2 + 1];        //传送显示数据
    while (!TI);
    TI = 0;
    SBUF = dispram[j][i * 2];            //传送显示数据
    while (!TI);
    }while (j);                          //完成一行数据的发送
G = 1;                                   //消隐（关闭显示）
P1 & = 0xf0;                             //行号端口清零
R_CLK = 1;                               //显示数据输入输出锁存器
P1 | = i;                                //写入行号
R_CLK = 0;                               //锁存显示数据
G = 0;                                   //打开显示
}
```

第 14 章

课程设计七:电子密码锁

14.1 系统功能

设计一个单片机控制的电子密码锁,要求能设定一组 6 位的数字开启密码并能存储,在进行开启操作时,如三次输入密码错误则进行鸣叫报警并锁死系统。如密码输入正确,则进行声光开启提示。

14.2 设计方案

本电子密码锁采用 C52 系列单片机作为控制器,显示器使用八个普通的七段共阳数码管,应用动态扫描完成信息显示;操作按键共设十二个,除 0～9 十个数字键外,还有两个密码设定与开锁操作键;报警与提醒电路采用喇叭与发光方式完成;存储器使用 AT24C01。总体的电路系统组成框图见图 14.1。程序软件采用 Keil-C51 编译器,用 C 语言编写控制代码。

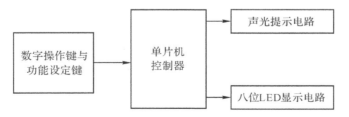

图 14.1 电子密码锁系统结构框图

14.3　系统硬件仿真电路

电子密码锁系统硬件电路组成大致上可以分成单片机控制电路与键盘电路、显示电路、存储电路、声光电路五部分。图 14.2 为硬件电路连接图。

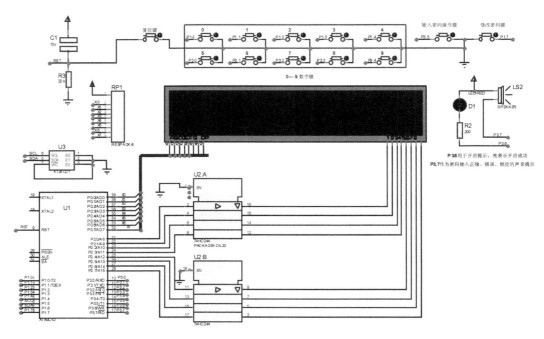

图 14.2　电子密码锁硬件电路连接图

14.3.1　单片机电路

单片机端口资源的分配主要有:P0 口负责输出显示段码,端口接上拉电阻;P2 口为 LED 扫描控制口,对每个数码管进行约 1ms 的轮流正电压供电;端口 P1.0～P1.4 分别对应数字键 0、1、2、3、4,端口 P3.0～P3.4 分别对应数字键 5、6、7、8、9;P3.5 端口为开启操作的功能键,开锁时先按一下开启操作键,然后再输入六个密码;P1.7 端口为修改密码操作键,当要更改密码时,先按一下修改密码键,然后输入六位的数字密码,当六位密码输入完毕,系统会自动将新密码存入存储器;P3.6 端口控制一个发光二极管用于锁开启提示,亮时表示开启成功,实际应用时可接电磁开锁线圈驱动电路;P3.7 端口连接一个扬声器作为密码输入正确及错误的声音提示,密码正确时响三响,前两次频率低,后一次频率高且时间较长。密码输入错误时响三次低频,可再次输入密码,但总共只有三次机会,三次密码错误后响六次较高频率的声响后自动锁死;P1.5 与 P1.6 端口分别为存储器

AT24C01 的时钟信号及数据信号端口。

14.3.2　键盘电路

键盘电路由于单片机端口数量足够直接使用了查询式按键电路,其中按键小开关的一端接单片机端口,另一端接地,当按键按下时,通过单片机查询确定哪个端口键按下了,然后执行相应的控制程序。图 14.3 为按键电路连接图,其中两个为功能键,十个为 0－9 数字键。

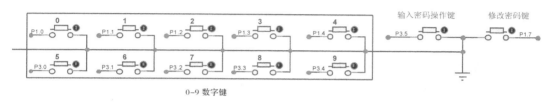

图 14.3　操作按键连接电路

14.3.3　数码管显示电路

LED 显示器采用共阳八位一体的显示器 7SEG－MPX8－CA－BLUE,其中左边的 8 个引脚分别对应段码输出最低位至最高位,DP 为小数点位。而右边的 8 条引脚为八位数码管的阳极供电端。图 14.4 为数码管显示器引脚图。

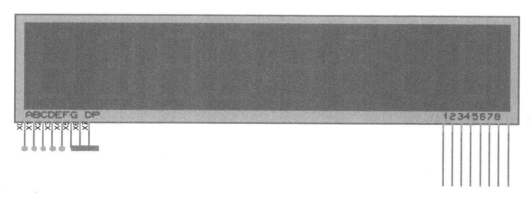

图 14.4　数码管显示器引脚图

14.3.4　密码存储电路

存储器 AT24C01 是容量为 1K 位的串行可电擦除只读存储器,可存储 128 个字节,该芯片支持 I^2C 总线数据传送协议,图 14.5 为 AT24C01 使用引脚连接图。

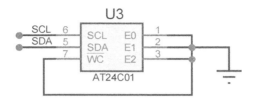

图 14.5　AT24C01 应用引脚连接图

14.3.5　声光提示电路

声音与光提示电路使用小喇叭与红色发光二极管。喇叭用单片机输出的方波信号驱动发声,在密码输入错误或正确时用不同频率的声音来区分。红色发光二极管亮时表示密码输入正确开锁成功。图 14.6 为喇叭与 LED 小灯提示电路。

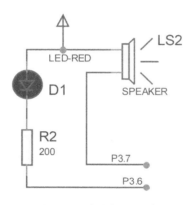

图 14.6　声光提示电路

14.4　程序设计

单片机电子密码锁控制程序采用 C 语言编写,使用 Keil-C51 编译器。控制程序模块主要有密码存取程序、开锁密码输入与比较程序、动态显示程序、发声程序等。

14.4.1　主程序

主程序控制的主要流程如图 14.7 所示。开机上电后先进行一些初始化工作,然后等待开锁命令或设定新密码命令。当按下设定新密码键时,可输入六位的数字密码,系统在输入六位数字后自动将新密码存入存储器,并在下次开机时自动先取出存储的密码放入内存作为比较用。当按下开锁键时,系统在输入六位数字后自动进行密码比较,如正确则声光提醒并开锁,如密码有误则发警告声三响并进入继续输入密码状态,当输入密码的错误次数达到三次时,发报警声响六响并锁住系统,使其不能进行任何操作。

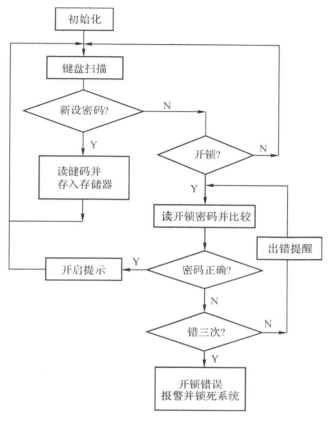

图 14.7　主程序工作流程图

14.4.2　初始化程序

初始化程序主要是对单片机的控制寄存器进行一些工作前的设定,规划用于数据处理的内存并将存储器中的密码读入内存用于开锁比较,调用显示程序使段码管全部发亮以检查发光管是否有损坏,最后进入主程序循环。

14.4.3　按键扫描程序

按键扫描主要是先对开锁键及密码重新设定键进行检查,如有按键按下则进入相应的操作流程,等待读入数字键的键值,并放入内存,当读入数字达到六个时进行存储或密码比较操作。

14.5　软件调试与运行结果

软件编程调试按从小到大、从易到难的方法，首先在 Proteus 软件中打开 ISIS 软件，画出仿真电路图，然后使用编译器 Keil-C51 进行程序的编写，主要编写与调试的模块程序次序如下：

（1）扫描显示程序。扫描显示程序主要是对显示缓存中的六个数据进行 LED 段码扫描显示，调试内容包括共阳段码表的段码是否正确、每个 LED 是否能点亮、左边与右边的次序是否正确等内容，图 14.8 为检查段码是否能全点亮的程序仿真运行图。

图 14.8　LED 显示六个"8"程序仿真图

（2）提示信息显示程序。提示信息显示方式大致有输入密码提示信息、出错提示信息、开启成功提示信息等，图 14.9 至图 14.14 为各类信息显示图。

图 14.9　等待输入新密码或等待输入开启密码状态图

图 14.10　输入新密码中状态图（图中已输入 4 位）

图 14.11　输入开启密码状态图（图中已输入 3 位）

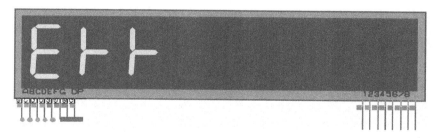

图 14.12　输入密码出错提示状态图

图 14.13　三次密码输入出错报警提示状态图

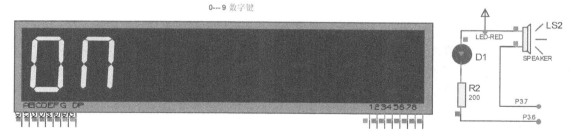

图 14.14　密码输入正确提示状态图(右边发光管亮表示开锁成功)

(3)提示信息发声程序。提示发声程序的编程原理是产生一定频率的方波信号输出给喇叭,电子密码锁程序中使用了三种方式的提示发声音,第一种是输入密码正确,为二低一高共三次的声响;第二种是输入密码错误,为三次低频声响;第三种是输入三次错误密码,为六次高频声响。

(4)存储器读写程序。根据 AT24C01 存储器器件手册及 I^2C 通信标准编写存储器读写程序,并将六个数据字节写入并读出进行比较,直到正确为止。

(5)读按键程序。按键程序采用顺序查键的方法,编程中应注意软件消抖以及按键等待释放等功能的使用。

(6)控制主程序。在以上各个子程序调试正确的基础上,可以按电子锁控制功能完成主程序的编写与调试,当然编程的方法也不是固定的,模块可以交叉编写或按自己的习惯编写。

14.6 C 源程序清单

```
//*******************************************************//
//                    密码开锁演示程序                    //
//                    2013 年 1 月 18 日                  //
//*******************************************************//
//
//**************** 预处理 ********************//
# include <reg52.h>
# include "24c01.h"          //存储器 24C01 读写程序
//
//**************** 定义 ********************//
sbit P10 = P1^0;
sbit P11 = P1^1;
sbit P12 = P1^2;
sbit P13 = P1^3;
sbit P14 = P1^4;
sbit P17 = P1^7;
sbit P30 = P3^0;
sbit P31 = P3^1;
sbit P32 = P3^2;
sbit P33 = P3^3;
sbit P34 = P3^4;
sbit P35 = P3^5;
sbit P36 = P3^6;
sbit P37 = P3^7;
//
char code dis_7[16] = {0xc0,0xf9,0xa4,0xb0,0x99,0x92,0x82,0xf8,0x80,0x90,0xbf,
0xc8,0xff,0xb6,0x86,0x8f};
// 共阳显示段码表： 0  1  2  3  4  5  6  7  8  9  -  n  不亮  三  E  r
char code scan_con[6] = {0x01,0x02,0x04,0x08,0x10,0x20};
//控制七段数码管的工作电压,0X01 表示最左端数码显示
char data dia[6] = {0x08,0x08,0x08,0x08,0x08,0x08};
//存储显示于数码管上的值,最开始显示"888888" 顺便检验数码管有没有损坏的现象
unsigned char data dat1[2][6] = {0,0,0,0,0,0};
//用于存储设定密码与输入密码数值;dat1[0][]为设定的密码 dat1[1][]为输入的密码
```

```
unsigned char data dat2[6];//显示缓存
bit flag = 0,flag1 = 0;
//
//***************** 函数声明 ******************//
//
void delayms(int t);
void scan(void);
void sound(char a, char b,char c);
void input(char x);
void displayon(void);
void displaywrong(void);
void displayagain(void);
void displayinput(void);
void jiemi(void);
//**************** 1ms 延时程序 ******************//
//
void delayms(int t)
{
    int i, j;
      for(i = 0;i<t;i++)
        for(j = 0;j<120;j++)
            ;
}
//*************** 显示扫描程序 ******************//
//
void scan(void)
{
    unsigned char k;
    for(k = 0;k<6;k++)
    {
      P0 = dis_7[dia[k]];
      P2 = scan_con[k];
      delayms(1);
      P0 = 0xff;
    }
}
//
//**************** 发声程序 ******************//
//a 表示响的时间,b 表示频率,c 表示响的次数
```

```
void sound(char a, char b,char c)
    {
        char i, j,k;
        for(k = 0;k<c;k ++ )
        {
            for(i = 0;i<a;i ++ )
                for(j = 0;j<100;j ++ )
                {
                    P37 = ! P37;
                    delayms(b);
                }
            delayms(700);
        }
    }
//
//************* 按键输入处理程序 *************************//
//
void input(char x)//参数 x = 0 表示输入设定值,x = 1 表示输入密码
{
char i, j = 0;//j 作为标志是否有按键输入的标志位
for(i = 0;i<6;i ++ ,j = 0)
while(j == 0)
{
  scan();
  if(P10 == 0)//10 端口表示设定 0
  {
    while(P10 == 0)    scan();
    if(x == 0)
        dia[i] = 13;
    else
        dia[i] = 0;
    j = 1;
    dat1[x][i] = 0;
  }
  if(P11 == 0)//11 端口表示设定 1
  {
    while(P11 == 0)    scan();
    if(x == 0)
        dia[i] = 13;
```

```
        else
            dia[i] = 1;
        j = 1;
        dat1[x][i] = 1;
    }
    if(P12 == 0)//12 端口表示设定 2
    {
        while(P12 == 0)    scan();
        if(x == 0)
            dia[i] = 13;
        else
            dia[i] = 2;
        j = 1;
        dat1[x][i] = 2;
    }
    if(P13 == 0)//13 端口表示设定 3
    {
        while(P13 == 0)    scan();
        if(x == 0)
            dia[i] = 13;
        else
            dia[i] = 3;
        j = 1;
        dat1[x][i] = 3;
    }
    if(P14 == 0)//14 端口表示设定 4
    {
        while(P14 == 0)    scan();
        if(x == 0)
            dia[i] = 13;
        else
            dia[i] = 4;
        j = 1;
        dat1[x][i] = 4;
    }
    if(P30 == 0)//30 端口表示设定 5
    {
        while(P30 == 0)    scan();
        if(x == 0)
```

```
                dia[i] = 13;
        else
                dia[i] = 5;
        j = 1;
        dat1[x][i] = 5;
}
if(P31 == 0)//31 端口表示设定 6
{
    while(P31 == 0)    scan();
    if(x == 0)
            dia[i] = 13;
    else
            dia[i] = 6;
    j = 1;
    dat1[x][i] = 6;
}
if(P32 == 0)//32 端口表示设定 7
{
    while(P32 == 0)    scan();
    if(x == 0)
            dia[i] = 13;
    else
            dia[i] = 7;
    j = 1;
    dat1[x][i] = 7;
}
if(P33 == 0)//33 端口表示设定 8
{
    while(P33 == 0)    scan();
    if(x == 0)
            dia[i] = 13;
    else
            dia[i] = 8;
    j = 1;
    dat1[x][i] = 8;
}
if(P34 == 0)//34 端口表示设定 9
{
    while(P34 == 0)    scan();
```

```
        if(x = = 0)
            dia[i] = 13;
        else
            dia[i] = 9;
        j = 1;
        dat1[x][i] = 9;
    }
  }
}
```

//****************** 开锁成功显示信息 ********************//
//
```
void displayon(void)//当密码正确时显示 on
{
dia[0] = 0x00; dia[1] = 0x0b;
dia[2] = 0x0c; dia[3] = 0x0c; dia[4] = 0x0c; dia[5] = 0x0c;
scan();
}
```

//**************** 密码输入三次错误后显示信息 ********************//
//当显示 ------ 时等待输入密码,当输入三次密码仍然错误显示 ErrErr
//
```
void displaywrong(void)
{
    dia[0] = 0x0e; dia[1] = 0x0f; dia[2] = 0x0f;
    dia[3] = 0x0e; dia[4] = 0x0f; dia[5] = 0x0f;
    scan();
}
```

//*************** 密码输入提示显示信息 ********************//
//当输入设定值之后显示 ------ 等待输入密码,显示 ------
//
```
void displayinput(void)
{
    dia[0] = 0x0a; dia[1] = 0x0a; dia[2] = 0x0a;
    dia[3] = 0x0a; dia[4] = 0x0a; dia[5] = 0x0a;
    scan();
}
```

//*************** 密码出错再输入提示程序 ********************//
//当密码输入错误,要求再次输入时先显示 Err,3 秒后显示 ------
//
```
void displayagain(void)
```

```
{
    int n;
    dia[0] = 0x0e; dia[1] = 0x0f; dia[2] = 0x0f;
    dia[3] = 0x0c; dia[4] = 0x0c; dia[5] = 0x0c;
//  以下显示 Err 约 3 秒
    for(n = 0;n<500;n++) scan();
//  以下显示 ------
    dia[0] = 0x0a; dia[1] = 0x0a; dia[2] = 0x0a;
    dia[3] = 0x0a; dia[4] = 0x0a; dia[5] = 0x0a;
    scan();
}
//***************** 密码核对程序 ********************//
//
void jiemi(void)
  {
    char i, j, m = 0;
    for(i = 0;i<3;i++)
    {
        input(1);
        for(j = 0;j<6;j++)
        {
        if(dat1[0][j]! = dat1[1][j])
        {
            m = 1;
                break;
        }
        }
    }
  if(m == 1)
  {
        if(i<2)
        {
        sound(3,3,3);//a 表示时间,b 表示频率,人耳辨别范围 20 到 20kHz,c 表示响的次数
        displayagain();
        m = 0;
        }
        else
        {
        sound(6,1,6);
        flag = 1;
```

```
                displaywrong();
            }
        }
    else if(m == 0)
    {
            sound(2,2,2);sound(5,1,1);
            P36 = 0;
            displayon();
            flag1 = 1;
            break;
        }
    }
}
//
//************** 主程序 *******************//
//
void main(void)
{
    //input(0);
    //delayms(1000);
    // jiemi();
    //sound(2,2,1);//b 越大,频率越低
    //sound(5,1,1);
    char i;
    init(); //初始化 AT24C01
    for(i = 0;i<6;i++)
    {
    //write_add(00 + i,dat1[0][i]);//在 23 地址处写入数据 0x55;
    //delay1(100);
        dat1[0][i] = read_add(00 + i);//读入 24C01 存储器中的密码数据(共 6 个)
        delay1(100);
    }
//
    while(1)
    {
    scan();//显示扫描程序
    if(P35 == 0)
    {
        if(flag == 0)
```

```
        {
            P36 = 1;
            displayinput();
            jiemi();
        }
    }
    if(P17 == 0)
    {
        while(P17 == 0) scan();
        if(flag == 0)
        {
            displayinput();    //显示 ------ 等待输入密码
            input(0);
            for(i = 0;i<6;i++)
            {
                write_add(00 + i,dat1[0][i]);//在 23 地址处写入数据 0x55;
                delay1(100);
                //dat1[0][i] = read_add(00 + i);用于测试看写入与读出的数据是否一样
                //delay1(100);
            }
            if(flag1 == 1)
                displayon();
            else
                displayinput();
        }
    }
}
//***************** 程序结束 *********************//
```

以下 AT24C01 头文件。

```
        //***********************************************************//
        //             AT24C01.H 头文件                              //
        //                2013 年 1 月 18 日                          //
        //***********************************************************//
//
# ifndef _24c01_H_
# define _24c01_H_
```

```
//
//***************** 预处理 ******************//
#define uchar unsigned char
//***************** 定义 ******************//
sbit sda = P1^6;
sbit scl = P1^5;
uchar a;
//***************** 延时程序 ******************//
void delay()
{ ;; }
//******************************************//
//开始信号
void start()
{
    sda = 1;
    delay();
    scl = 1;
    delay();
    sda = 0;
    delay();
}
//******************************************//
//停止
void stop()
{
    sda = 0;
    delay();
    scl = 1;
    delay();
    sda = 1;
    delay();
}
//******************************************//
//应答,在数据传送 8 位后,等待或者发送一个应答信号
void respons()
{
    uchar i;
    scl = 1;
    delay();
```

```
        while((sda == 1)&&(i<250))i++;
        scl = 0;
        delay();
}
//************************************************//
//初始化函数,拉高 sda 和 scl 两条总线
void init()
{
        sda = 1;
        scl = 1;
}
/***********************************************************************/
/* 根据数据有效性规则,读写数据时必须将 SCL 拉高,然后送入或读出数据,完毕后再将
SCL 拉低 */
/***********************************************************************/
//写一字节,将 date 写入 AT24C01 中
void write_byte(uchar date)
{
        uchar i;
        scl = 0;
        for(i = 0;i<8;i++)
        {
          date = date<<1;
          sda = CY; //将要送入数据送入 sda
          scl = 1; //scl 拉高准备写数据
          delay();
          scl = 0; //scl 拉低数据写完毕
          delay();
        }
}
uchar read_byte()//读取一字节,从 AT24C02 中读取一个字节
{
        uchar i,k;
        for(i = 0;i<8;i++)
        {
          scl = 1; //scl 拉高准备读数据
          delay();
          k = (k<<1)|sda; //将 sda 中的数据读出
          scl = 0; //scl 拉低数据写完毕
```

```
            delay();
        }
        return k;
}
//************************************//
//延时程序
void delay1(uchar x)
{
    uchar a,b;
    for(a = x;a>0;a--)
    for(b = 100;b>0;b--);
}
/******************************************************************/
/* 读出与写入数据时必须严格遵守时序要求 */
/******************************************************************/
//向 AT24C01 中写数据
void write_add(uchar address,uchar date)
{
    start();
    write_byte(0xa0);
    respons();
    write_byte(address);
    respons();
    write_byte(date);
    respons();
    stop();
}
//************************************//
//从 AT24C01 中读出数据
uchar read_add(uchar address)
{
    uchar date;
    start();
    write_byte(0xa0);
    respons();
    write_byte(address);
    respons();
    start();
    write_byte(0xa1);
```

```
        respons();
        date = read_byte();
        stop();
        return date;
    }
    #endif
    //
    //************ **** AT24C01 头文件结束 ******************//
```

附 录 1

51 系列单片机的特殊功能寄存器表

符号	寄存器名	位地址、位标记及位功能								直接地址	复位状态
		D7	D6	D5	D4	D3	D2	D1	D0		
(1)可位寻址 SFR(11)											
ACC	累加器	E7	E6	E5	E4	E3	E2	E1	E0	E0H	00H
		ACC. 7	ACC. 6	ACC. 5	ACC. 4	ACC. 3	ACC. 2	ACC. 1	ACC. 0		
B	B 寄存器	F7	F6	F5	F4	F3	F2	F1	F0	F0H	00H
		B. 7	B. 6	B. 5	B. 4	B. 3	B. 2	B. 1	B. 0		
PSW	程序状态字	D7	D6	D5	D4	D3	D2	D1	D0	D0H	00H
		CY	AC	F0	RS1	RS0	OV	—	P		
IP	中断优先权寄存器	BF	BE	BD	BC	BB	BA	B9	B8	B8H	×××00000B
		—	—	—	PS	PT1	PX1	PT0	PX0		
P3	P3 口	B7	B6	B5	B4	B3	B2	B1	B0	B0H	FFH
		P3. 7	P3. 6	P3. 5	P3. 4	P3. 3	P3. 2	P3. 1	P3. 0		
IE	中断允许寄存器	AF	AE	AD	AC	AB	AA	A9	A8	A8H	0××00000B
		EA	—	—	ES	ET1	EX1	ET0	EX0		
P2	P2 口	A7	A6	A5	A4	A3	A2	A1	A0	A0H	FFH
		P2. 7	P2. 6	P2. 5	P2. 4	P2. 3	P2. 2	P2. 1	P2. 0		
SCON	串行口控制寄存器	9F	9E	9D	9C	9B	9A	99	98	98H	00H
		SM0	SM1	SM2	REN	TB8	RB8	TI	RI		
P1	P1 口	97	96	95	94	93	92	91	90	90H	FFH
		P1. 7	P1. 6	P1. 5	P1. 4	P1. 3	P1. 2	P1. 1	P1. 0		
TCON	定时器控制寄存器	8F	8E	8D	8C	8B	8A	89	88	88H	00H
		TF1	TR1	TF0	TR0	IE1	IT1	IE0	IT0		
P0	P0 口	87	86	85	84	83	82	81	80	80H	FFH
		P0. 7	P0. 6	P0. 5	P0. 4	P0. 3	P0. 2	P0. 1	P0. 0		

符号	寄存器名	位地址、位标记及位功能								直接地址	复位状态
		D7	D6	D5	D4	D3	D2	D1	D0		
(2)不可位寻址 SFR(10)											
SP	栈指示器									81H	07H
DPL	数据指针 低 8 位									82H	00H
DPH	数据指针 高 8 位									83H	00H
PCON	电源控制 寄存器	SMOD	⋯	⋯	⋯	GF1	GF0	PD	IDL	87H	0×××0000B
TMOD	定时器方式 寄存器	GATE	C/$\overline{\text{T}}$	M1	M0	GATE	C/$\overline{\text{T}}$	M1	M0	89H	00H
TL0	T0 寄存器 低 8 位									8AH	00H
TL1	T1 寄存器 低 8 位									8BH	00H
TH0	T0 寄存器 高 8 位									8CH	00H
TH1	T1 寄存器 高 8 位									8DH	00H
SBUF	串行口数据 缓冲器									99H	×××× ××××B

附 录 2

51 系列单片机中断入口地址表

ROM 地址	用　　　途	优先级
0000H	复位程序运行入口	
0003H	外中断 0 入口地址	
000BH	定时器 T0 溢出中断入口地址	
0013H	外中断 1 入口地址	↓
001BH	定时器 T1 溢出中断入口地址	
0023H	串行口发送/接收中断入口地址	

附 录 3

51 系列单片机汇编指令表

1. 数据传送指令（29 条）

汇编指令	操作说明	代码长度 字节	指令周期	
			T_{ose}	T_m
（1）程序存储器查表指令 2 条				
MOVC A,@A+DPTR	将以 DPTR 为基址,A 为偏移地址中的数送入 A 中	1	24	2
MOVC A,@A+PC	将以 PC 为基址,A 为偏移地址中的数送入 A 中	1	24	2
（2）片外 RAM 传送指令 4 条				
MOVX A,@DPTR	将片外 RAM 中的 DPTR 地址中的数送入 A 中	1	24	2
MOVX @DPTR,A	将 A 中的数送入片外 RAM 中的 DPTR 地址单元中	1	24	2
MOVX A,@Ri	将片外 RAM 中@Ri 指示的地址中的数送入 A 中	1	24	2
MOVX @Ri,A	将 A 中的数送入片外@Ri 指示的地址单元中	1	24	2
（3）片内 RAM 及寄存器间数据传送指令（18 条）				
MOV A,Rn	将 Rn 中的数送入 A 中	1	12	1
MOV A,direct	将直接地址 direct 中的数送入 A 中	2	12	1
MOV A,#data	将 8 位常数送入 A 中	2	12	1
MOV A,@Ri	将 Ri 指示的地址中的数送入 A 中	1	12	1
MOV Rn,direct	将直接地址 direct 中的数送入 Rn 中	2	24	2
MOV Rn,#data	将立即数送入 Rn 中	2	12	1
MOV Rn,A	将 A 中的数送入 Rn 中	1	12	1
MOV direct,Rn	将 Rn 中的数送入 direct 中	2	24	2
MOV direct,A	将 A 中的数送入 direct 中	2	12	1
MOV direct,@Ri	将@Ri 指示单元中的数送入 direct 中	2	24	2

续表

汇编指令	操作说明	代码长度字节	指令周期	
			Tosc	Tm
MOV　direct,#data	将立即数送入 direct 中	3	24	2
MOV　direct,direct	将一个 direct 中的数送入另一个 direct 中	3	24	2
MOV　@Ri,A	将 A 中的数送入 Ri 指示的地址中	1	12	1
MOV　@Ri,direct	将 direct 中的数送入 Ri 指示的地址中	2	24	2
MOV　@Ri,#data	将立即数送入 Ri 指示的地址中	2	12	1
MOV　DPTR,#data16	将 16 位立即数直接送入 DPTR 中	3	24	2
PUSH　direct	将 direct 中的数压入堆栈	2	24	2
POP　direct	将堆栈中的数弹出到 direct 中	2	24	2
（4）数据交换指令（5 条）				
XCH　A,Rn	A 中的数和 Rn 中数全交换	1	12	1
XCH　A,direct	A 中的数和 direct 中数全交换	2	12	1
XCH　A,@Ri	A 中的数和@Ri 中数全交换	1	12	1
XCHD　A,@Ri	A 中的数和@Ri 中数半交换	1	12	1
SWAP　A	A 中数自交换（高 4 位与低 4 位）	1	12	1

2. 算术运算类指令（24 条）

汇编指令	操作说明	代码长度字节	指令周期	
			Tosc	Tm
ADD　A,Rn	Rn 中与 A 中的数相加,结果在 A 中,影响 PSW 位的状态	1	12	1
ADD　A,direct	direct 中与 A 中的数相加,结果在 A 中,影响 PSW 位的状态	2	12	1
ADD　A,#data	立即数与 A 中的数相加,结果在 A 中,影响 PSW 位的状态	2	12	1
ADD　A,@Ri	@Ri 中与 A 中的数相加,结果在 A 中,影响 PSW 位的状态	1	12	1
ADDC　A,Rn	Rn 中与 A 中的数带进位加,结果在 A 中,影响 PSW 位的状态	1	12	1
ADDC　A,direct	direct 中与 A 中的数带进位加,结果在 A 中,影响 PSW 位的状态	2	12	1
ADDC　A,#data	立即数与 A 中的数带进位加,结果在 A 中,影响 PSW 位的状态	2	12	1
ADDC　A,@Ri	@Ri 中与 A 中的数带进位加,结果在 A 中,影响 PSW 位的状态	1	12	1
SUBB　A,Rn	Rn 中与 A 中的数带借位减,结果在 A 中,影响 PSW 位的状态	1	12	1

汇编指令	操作说明	代码长度 字节	指令周期	
			Tosc	Tm
SUBB　A,direct	direct 中与 A 中的数带借位减,结果在 A 中,影响 PSW 位的状态	2	12	1
SUBB　A,♯data	立即数与 A 中的数带借位减,结果在 A 中,影响 PSW 位的状态	2	12	1
SUBB　A,@Ri	@Ri 中与 A 中的数带借位减,结果在 A 中,影响 PSW 位的状态	1	12	1
INC　A	A 中数加 1	1	12	1
INC　Rn	Rn 中数加 1	1	12	1
INC　direct	direct 中数加 1	2	12	1
INC　@Ri	@Ri 中数加 1	1	12	1
INC　DPTR	DPTR 中数加 1	1	24	2
DEC　A	A 中数减 1	1	12	1
DEC　Rn	Rn 中数减 1	1	12	1
DEC　direct	direct 中数减 1	2	12	1
DEC　@Ri	@Ri 中数减 1	1	12	1
MUL　AB	A、B 中两无符号数相乘,结果低 8 位在 A 中,高 8 位在 B 中	1	48	4
DIV　AB	A、B 中两无符号数相除,商在 A 中,余数在 B 中	1	48	4
DA　A	十进制调整,对 BCD 码十进制加法运算结果调整(不适合减法)	1	12	1

3. 逻辑运算指令(24 条)

汇编指令	操作说明	代码长度 字节	指令周期	
			Tosc	Tm
ANL　A,Rn	Rn 中与 A 中的数相"与",结果在 A 中	1	12	1
ANL　A,direct	direct 中与 A 中的数相"与",结果在 A 中	2	12	1
ANL　A,♯data	立即数与 A 中的数相"与",结果在 A 中	2	12	1
ANL　A,@Ri	@Ri 中与 A 中的数相"与",结果在 A 中	1	12	1
ANL　direct,A	A 和 direct 中数进行"与"操作,结果在 direct 中	2	12	1
ANL　direct,♯data	常数和 direct 中数进行"与"操作,结果在 direct 中	3	24	2
ORL　A,Rn	Rn 中和 A 中数进行"或"操作,结果在 A 中	1	12	1
ORL　A,direct	direct 中和 A 中数进行"或"操作,结果在 A 中	2	12	1

续表

汇编指令	操作说明	代码长度字节	指令周期 Tosc	指令周期 Tm
ORL A, #data	立即数和 A 中数进行"或"操作,结果在 A 中	2	12	1
ORL A,@Ri	@Ri 中和 A 中数进行"或"操作,结果在 A 中	1	12	1
ORL direct,A	A 中和 direct 中数进行"或"操作,结果在 direct 中	2	12	1
ORL direct, #data	立即数和 direct 中数进行"或"操作,结果在 direct 中	3	24	2
XRL A,Rn	Rn 中和 A 中数进行"异或"操作,结果在 A 中	1	12	1
XRL A,direct	direct 中和 A 中数进行"异或"操作,结果在 A 中	2	12	1
XRL A, #data	立即数和 A 中数进行"异或"操作,结果在 A 中	2	12	1
XRL A,@Ri	@Ri 中和 A 中数进行"异或"操作,结果在 A 中	1	12	1
XRL direct,A	A 中和 direct 中数进行"异或"操作,结果在 direct 中	2	12	1
XRL direct, #data	立即数和 direct 中数进行"异或"操作,结果在 direct 中	3	24	2
RR A	A 中数循环右移(移向低位)D0 移入 D7	1	12	1
RRC A	A 中数带进位循环右移(D0 移入 C,C 移入 D7)	1	12	1
RL A	A 中数循环左移(移向高位)D7 移入 D0	1	12	1
RLC A	A 中数带进位循环左移(D7 移入 C,C 移入 D0)	1	12	1
CLR A	A 中数清零	1	12	1
CPL A	A 中数取反	1	12	1

4. 程序转移类指令(17 条)

汇编指令	操作说明	代码长度字节	指令周期 Tosc	指令周期 Tm
(1)无条件转移指令 9 条				
LJMP addr16	长转移,程序转到 addr16 指示的地址处	3	24	2
AJMP addr11	短转移,程序转到 addr11 指示的地址处	2	24	2
SJMP rel	相对转移,程序转到 rel 指示的地址处	2	24	2
LCALL addr16	长调用,程序调用 addr16 处的子程序	3	24	2

汇编指令	操作说明	代码长度字节	指令周期	
			Tosc	Tm
ACALL addr11	短调用,程序调用 addr11 处的子程序	2	24	2
JMP @A+DPTR	程序散转,程序转到 DPTR 为基址,A 为偏移地址处	1	24	2
RETI	中断返回	1	24	2
RET	子程序返回	1	24	2
NOP	空操作	1	12	1
(2)条件转移指令 8 条				
JZ rel	A 中数为零,程序转到相对地址 rel 处	2	24	2
JNZ rel	A 中数不为零,程序转到相对地址 rel 处	2	24	2
DJNZ Rn,rel	Rn 中数减 1 不为零,程序转到相对地址 rel 处	2	24	2
DJNZ direct,rel	direct 中数减 1 不为零,程序转到相对地址 rel 处	3	24	2
CJNE A,♯data,rel	♯data 与 A 中数不等转至 rel 处。C=1,data＞(A);C=0,data＜=(A)	3	24	2
CJNE A,direct,rel	direct 与 A 中数不等转至 rel 处。C=1,data＞(A);C=0,data＜=(A)	3	24	2
CJNE Rn,♯data,rel	♯data 与 Rn 中数不等转至 rel 处。C=1,data＞(Rn);C=0,data＜=(Rn)	3	24	2
CJNE @Ri,♯data,rel	♯data 与 @Ri 中数不等转至 rel 处。C=1,data＞(@Ri);C=0,data＜=((@Ri)	3	24	2

5. 布尔指令(17 条)

汇编指令	操作说明	代码长度字节	指令周期	
			Tosc	Tm
(1)位操作指令(12 条)				
MOV C,bit	bit 中状态送入 C 中	2	12	1
MOV bit,C	C 中状态送入 bit 中	2	24	2
ANL C,bit	bit 中状态与 C 中状态相"与",结果在 C 中	2	24	2
ANL C,/bit	bit 中状态取反与 C 中状态相"与",结果在 C 中	2	24	2
ORL C,bit	bit 中状态与 C 中状态相"或",结果在 C 中	2	24	2
ORL C,/bit	bit 中状态取反与 C 中状态相"或",结果在 C 中	2	24	2
CLR C	C 中状态清零	1	12	1
SETB C	C 状态置 1	1	12	1

续表

汇编指令	操作说明	代码长度字节	指令周期	
			Tosc	Tm
CPL C	C 中状态取反	1	12	1
CLR bit	bit 中状态清零	2	12	1
SETB bit	bit 中状态置1	2	12	1
CPL bit	bit 中状态取反	2	12	1
(2)位条件转移指令(5条)				
JC rel	进位位为1时,程序转至 rel	2	24	2
JNC rel	进位位不为1时,程序转至 rel	2	24	2
JB bit,rel	bit 状态为1时,程序转至 rel	3	24	2
JNB bit,rel	bit 状态不为1时,程序转至 rel	3	24	2
JBC bit,rel	bit 状态为1时,程序转至 rel,同时 bit 位清零	3	24	2

参考文献

1. 楼然苗,李光飞. 51系列单片机设计实例.北京:北京航空航天大学出版社,2003.

2. 楼然苗,李光飞. 51系列单片机设计实例(第2版). 北京:北京航空航天大学出版社,2005.

3. 楼然苗,李光飞.单片机课程设计指导. 北京:北京航空航天大学出版社,2007.

4. 李光飞,楼然苗,胡佳文等.单片机课程设计实例指导. 北京:北京航空航天大学出版社,2004.

5. 李光飞,李良儿,楼然苗等.单片机C程序设计实例指导. 北京:北京航空航天大学出版社,2005.

6. 楼然苗,胡佳文,李光飞等.51系列单片机原理及设计实例. 北京:北京航空航天大学出版社,2010.

7. 楼然苗,李光飞编著.单片机课程设计指导(第2版). 北京:北京航空航天大学出版社,2012.